Nawzt Sadiq Jaafar Al-Bayati
Raid Rafi Omar Al-Nima

Alzheimer's Disease Classification

Nawzt Sadiq Jaafar Al-Bayati
Raid Rafi Omar Al-Nima

Alzheimer's Disease Classification

by Employing Various Intelligent Techniques

Noor Publishing

Cover image: www.ingimage.com

Publisher:
Noor Publishing
is a trademark of
Dodo Books Indian Ocean Ltd. and OmniScriptum S.R.L publishing group

120 High Road, East Finchley, London, N2 9ED, United Kingdom
Str. Armeneasca 28/1, office 1, Chisinau MD-2012, Republic of Moldova, Europe
Managing Directors: Ieva Konstantinova, Victoria Ursu
info@omniscriptum.com

Printed at: see last page
ISBN: 978-620-9-52641-1

Alzheimer's Disease Classification by Employing Various Intelligent Techniques

Nawzt Sadiq Jaafar Al-Bayati

Assist. Prof. Dr. Raid Rafi Omar Al-Nima

ABSTRACT

One of big medical issues which is progressive and prevalent is known as Alzheimer's Disease (AD). It is related to biochemical and structural alterations in a brain, where it leads to the gathering certain proteins. Such alterations cause eventual destruction and progressive shrinkage for brain cells, highlighting AD as a significant neurodegenerative medical issue.

This book suggests hybrid Artificial Intelligence (AI) approaches. Each one of them can optimize a Convolutional Neural Network (CNN) architecture with its hyperparameters. The used approaches are called: Optimized CNN by Optuna (OCNNO) and Optimized CNN by Keras-Tuner (OCNNKT). AD system is also suggested and implemented. It utilizes software and hardware components, like Python and Raspberry Pi 5 (RP5). The system has the ability to diagnose (or classify) AD into its level of: Very Mild Dementia (VMD), Moderate Dementia (MoD), Mild Dementia (MD) and Non-Dementia (ND). It has other facilitates as providing quick assessment and useful communication with healthcare professionals.

Two datasets are considered for implementations. They are termed the: Best Alzheimer Magnetic Resonance Imaging (BA-MRI) and Alzheimer's Disease Magnetic Resonance Imaging (AD-MRI). Noticeably, comparable and excellent outcomes are achieved. The OCNNO evaluation accuracies of 99.18% and 95.79% have been attained for the BA-MRI and AD-MRI datasets, respectively. The OCNNKT testing accuracies of 93.17% and 92.44% have been obtained for the BA-MRI and AD-MRI datasets, respectively.

TABLE OF CONTENTS

LIST OF FIGURES

LIST OF TABLES

LIST OF ABRREVIATIONS

Abbreviation	Meaning
AD	Alzheimer’s Disease
AD-MRI	Alzheimer’s Disease Magnetic Resonance Imaging
ADNI	Alzheimer Disease Neuroimaging Initiative
AI	Artificial Intelligence
Alex-Net	Alex Network
ANFIS	Adaptive Neuro Fuzzy Inference System
ANN	Artificial Neural Network
BA	Beta-Amyloid
BA-MRI	Best Alzheimer Magnetic Resonance Imaging
BO	Bayesian Optimization
BoVW	Bag-of- Visual Words
CAD	Computer-Aided Diagnosis
CF	Cerebrospinal Fluid
CMA-ES	Covariance Matrix Adaptation-Evolution Strategy
CNN	Convolutional Neural Network
CvT	Curvelet Transform
DANN	Domain Adversarial Neural Network
DCNN	Deep Convolutional Neural Network

Abbreviation	Meaning
DenseNet121	Densely connected convolutional Network-121
DL	Deep Learning
DML	Deep Metric Learning
DNN	Deep Neural Network
DT	Decision Tree
DTI	Diffusion Tensor Imaging
EEG	Electroencephalogram
ERPNet	Extra Event-Related Potential Network
FA	Fractional Anisotropy
FC	Fully Connected
fMRI	Functional Magnetic Resonance Imaging
FN	False Negative
FP	False Positive
FS	Fisher Score
GEO	Gene Expression Omnibus
GL-CHFs	Gauss-Laguerre Harmonic Functions
GLCM	Gray Level Co-occurrence Matrix
GM	Gray Matter
GUI	Graphical User Interference

Abbreviation	Meaning
HCV	Hippocampal Volume
HDMI	High-Definition Multimedia Interface
HT	Hyperband Technology
JPG	Joint Photographic Group
K-NN	K-Nearest Neighbors
LBD	Lewy Body Dementia
LRM	Logistic Regression Model
LVQ	Learning Vector Quantization
MD	Mild Dementia
MMSE	Mini-Mental State Examination
MoD	Moderate Dementia
NB	Naïve Bayes
ND	Non-Dementia
NIfTI	Neuroimaging Informatics Technology Initiative
OASIS	Open-Source Imaging Studies Series
OCNNKT	Optimized CNN by Keras-Tuner Optimization
OCNNO	Optimized CNN by Optuna Optimization
PCA	Principal Component Analysis
PDF	Portable Document Format

Abbreviation	Meaning
PET	Positron Emission Tomography
PRN	Pattern Recognition Network
PSEN1	PRESENILIN-1 gene
PSEN2	PRESENILIN-2 gene
PZM	Pseudo Zernike Moments
RAM	Randum Access Memory
ReLU	Rectified Linear Unit
ResNet50	Residual Network of 50 layers
ROI	Region of Interest
RP5	Raspberry Pi 5
RS	Random Search
SVM	Support Vector Machine
TL	Transfer Learning
TM	Trademark
TN	True Negative
TP	True Positive
TPE	Tree-structured Parzen Estimator
T-Tau	Total Tau
VAE	Variational Autoencoder

Abbreviation	Meaning
VGG16	Visual Geometry Group-16
VGG19	Visual Geometry Group-19
VMD	Very Mild Dementia
WHO	World Health Organization
WM	White Matter
XAM	X-axis Classification Model
YCM	Y-axis Classification Model
ZCM	Z-axis Classification Model

CHAPTER ONE

INTRODUCTION

1.1. Human's Brain:

Human's brain is often considered the most intricate and significant organ in the body [1]. This is due to the fact that it serves as the command and control center for all cognitive and biological functions. The brain is comprised of billions of neurons that interact with one another via the use of intricate chemical and electrical impulses [2]. As a result of these interactions, processes as thinking, memory, learning, movement, and emotions are brought about. The brain is composed of three primary regions: the cerebrum, the cerebellum and the brainstem. Each of these regions is responsible for a specialized function that is essential to the functioning of the body. In order to handle the messages that are transmitted between the organs of the body and the outside world, the brain collaborates with the nervous system to accomplish this task [3]. Figure 1.1 illustrates the main parts of the brain [4].

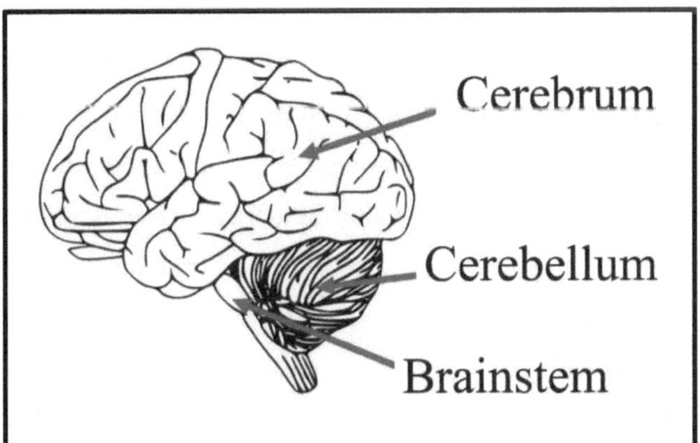

Figure 1.1: Demonstration of human's brain and its main parts

The brain is responsible for processing sensory information, such as hearing and vision Additionally, it controls various functions, including decision-making and motor activities [5].

In spite of its significance, the brain is vulnerable to a wide variety of disorders that have the potential to disrupt its activities. Conditions such as stroke, epilepsy, Parkinson's disease and multiple sclerosis present significant challenges to neurological health. Among these, AD stands out as one of the most prevalent neurodegenerative disorders, profoundly affecting memory and cognitive abilities [6].

1.2. Brain Diseases Overview:

Human's brain is susceptible to many diseases commonly known as brain diseases. These diseases include a wide range of disorders that affect the structure or function of the brain, resulting on the impacts in an individual's physical, cognitive and emotional health. The repercussions of these diseases are not limited to the affected individuals, they are also extended to families and communities, impacting quality of life and productivity [7].

Brain diseases are classified into several major groups [8]. Neurodegenerative diseases such as AD and Parkinson's disease, which are characterized by the progressive deterioration of neurons and their impact on cognitive and motor functions. Cerebrovascular diseases, such as strokes and cerebral aneurysms, result from disturbances in blood flow to the brain and often have devastating and rapid effects. Infectious diseases such as meningitis and encephalitis are caused by pathogens that damage brain tissue and pose a serious risk if not treated appropriately. In addition, traumatic brain injuries, usually caused by accidents, can lead to long-term complications [8]. Next, most commonly occurring brain diseases will be defined and illustrated.

1.2.1. Cerebral aneurysm:

Cerebral aneurysm is identified as a severe medical condition that arises when the walls of the blood vessels within the brain being weak, resulting in a balloon-like bulge or sac [9]. This condition is often asymptomatic during its early levels, presenting a considerable challenge in terms of diagnosis. However, when an aneurysm ruptures, the onset of acute symptoms can occur rapidly, often resulting in severe clinical manifestations, including sudden and severe headache, nausea, visual disturbances and loss of consciousness [9]. Figure 1.2 illustrates the anatomical representation of cerebral aneurysm [10].

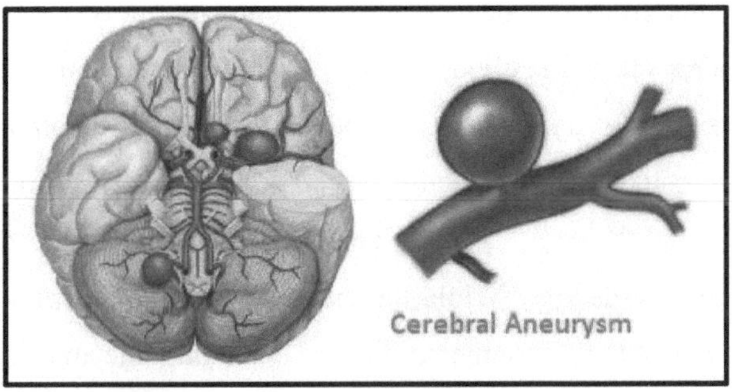

Figure 1.2: Anatomical representation of cerebral aneurysm, the left subfigure shows brain's vascular with multiple aneurysms and the right subfigure provides a magnified view of a cerebral aneurysm

The causes of cerebral aneurysms are diverse. Genetic predispositions are recognized as significant contributors here, alongside other factors such as hypertension and atherosclerosis [11]. Additional causes include head trauma and certain hereditary disorders, which may exacerbate the likelihood of developing a cerebral aneurysm [11].

Cerebral aneurysms are diagnosed using imaging techniques such as MRI and Computed Tomography (CT) scans, which identify their sizes and locations. Treatment varies based on the aneurysm's characteristics and it may involve monitoring or surgical repair. Preventative measures, including blood pressure control, smoking cessation and balanced diet are critical in reducing the risk of developing cerebral aneurysm. Prompt medical intervention is vital to prevent life-threatening complications in the case of rupture.

1.2.2. Parkinson's disease:

Parkinson's disease is a chronic and progressive neurological disorder named based on Dr. James Parkinson, who first described it in 1817. It results from the degeneration of dopamine-producing neurons in the brain's substantia nigra, leading to decreased dopamine levels that disrupt smooth and control muscle movements [12]. Symptoms include motor issues like tremors, rigidity, bradykinesia and postural instability, and non-motor issues such as depression, anxiety, sleep disturbances and cognitive decline. While the exact cause is unknown, a combination of genetic and environmental factors, including age and exposure to toxins, are believed to increase the risk of developing Parkinson's disease.

Diagnosing Parkinson's disease involves medical history reviews and physical exams, as no definitive test exists. Treatment focuses on symptom management through medications like Levodopa, therapies and, in advanced cases, surgical options like deep brain stimulation. Managing the disease requires a multidisciplinary approach, incorporating regular exercise, balanced diet and stress management.

Figure 1.3 provides a comparative between a healthy part of brain and a Parkinson's affected part of brain. It highlights structures such as the substantia nigra, red nucleus, reticular formation, cerebral aqueduct and superior colliculus. A significant focus is on the substantia nigra, which is visibly altered in the Parkinson's brain. This region is essential for producing dopamine, a neurotransmitter critical for smoothing and coordinating muscle movements [13].

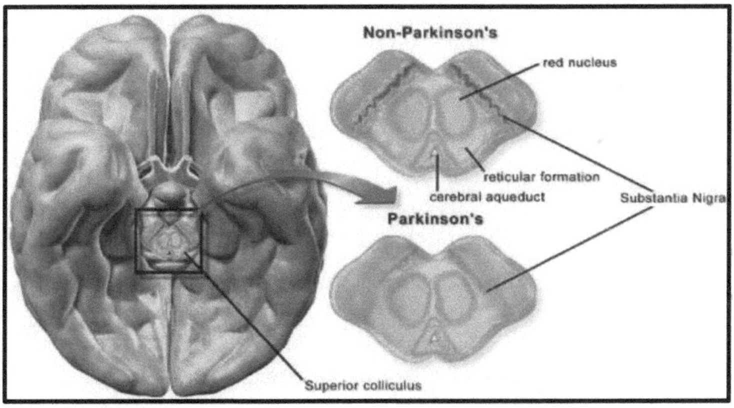

Figure 1.3: Structural differences between a healthy part of brain and a Parkinson's affected part of brain

1.2.3. Brain tumor:

Brain tumors are abnormal growths of cells within or around the brain that can significantly disrupt its normal function, and they are classified as benign or malignant based on their nature and behavior. Benign tumors grow slowly and are less likely to spread, whereas, malignant tumors are aggressive and can invade nearby tissues [14]. These tumors can either originate in the brain (primary tumors) or spread from other parts of the body (secondary tumors). Symptoms vary depending on the tumor's size, location and growth rate, including persistent headaches, nausea, seizures, vision or hearing problems, balance issues, and cognitive or personality changes. While the exact causes remain unclear, genetic mutations and environmental factors are believed to contribute with risk factors such as family history, radiation exposure and age. The diagnosis of brain tumors typically requires a combination of neurological assessments, advanced imaging techniques such as MRI or CT scans, and, in certain cases, a biopsy to confirm the nature of the tumor [15]. Treatment options include surgery to remove the tumor, radiation therapy, chemotherapy and targeted drug therapies, all of which aim to manage symptoms and slow tumor growth [15]. Rehabilitation therapies, emotional support, and multidisciplinary care are essential for helping patients who navigate the challenges of living with brain tumors.

Figure 1.4 illustrates a glioblastoma, an aggressive type of brain tumor, which is located at the upper left region of a human's brain. It visually highlights the tumor's appearance and position, making it an effective tool for understanding its impact on brain anatomy [16].

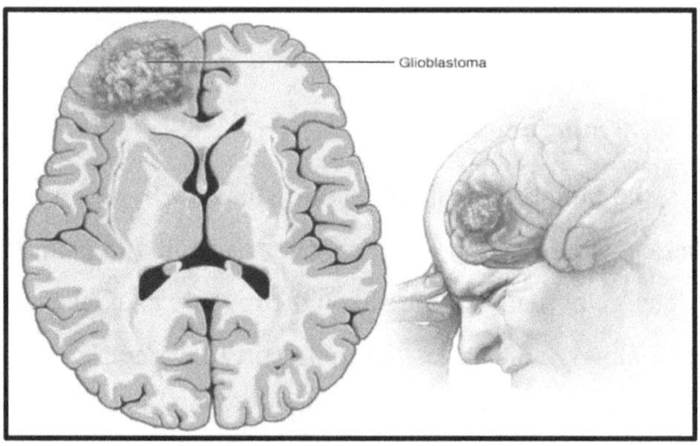

Figure 1.4: Anatomical representation of a glioblastoma in a human's brain

1.2.4. Vascular dementia:

Vascular dementia is a form of dementia caused by reduced blood flow to the brain, which leads to cognitive decline and affects reasoning, memory and judgment.

It is the second most common type of dementia after AD and often results from strokes, transient ischemic attacks or other vascular issues that damage brain tissue due to a lack of oxygen and nutrients [17]. Symptoms depend on the affected brain regions and they include confusion, difficulty concentrating, slowed thinking, problems with organization and decision-making, mood changes, memory loss, and unsteady gait. Risk factors such as high blood pressure, diabetes, high cholesterol, smoking, age, and history of heart disease or stroke increase the likelihood of developing vascular dementia [18].

Diagnosis typically involves neurological exams, imaging tests like MRI or CT scans, and assessments of cognitive functions.

Treatment focuses on managing underlying conditions through medications, lifestyle changes, and therapies to prevent further brain damage and improve symptoms.

Figure 1.5 illustrates the primary classifications of stroke, encompassing hemorrhagic stroke, which results from internal bleeding due to the rupture of a blood vessel, and ischemic stroke, which arises from an arterial blockage that disrupts cerebral blood flow. both types contribute to neuronal damage, thereby elevating the risk of vascular dementia [19].

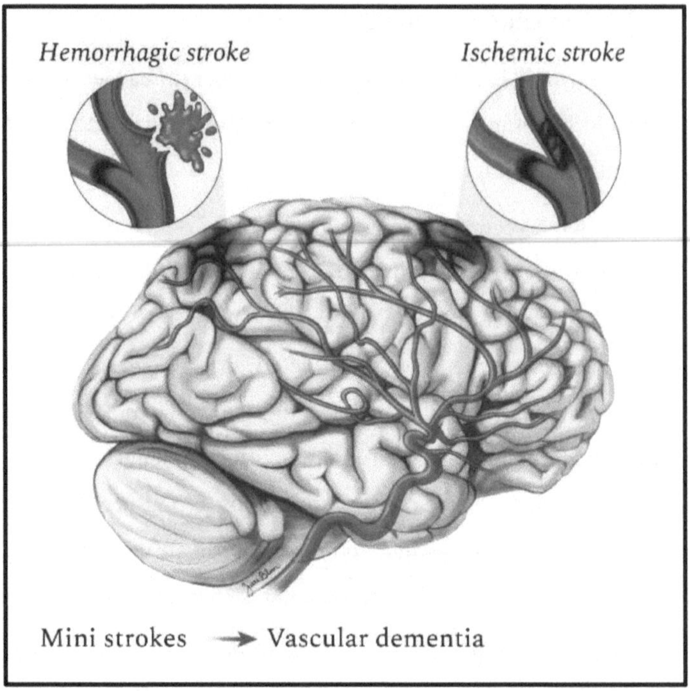

Figure 1.5: Stroke classification and its association with vascular dementia

1.2.5. Lewy Body Dementia (LBD):

LBD is a progressive neurological disorder caused by the accumulation of abnormal protein formations known as Lewy bodies in the brain, which disrupt neuronal function and lead to cognitive, behavioral, and motor impairments [20]. Cognitive symptoms include confusion, memory loss, poor attention and difficulty with visual-spatial tasks, while behavioral symptoms such as hallucinations, depression and apathy are also common. Motor symptoms, resembling those of Parkinson's, include muscle rigidity, tremors, slowed movements, and walking difficulties, along with sleep disturbances and autonomic dysfunction affecting blood pressure, digestion, and heart rate. The exact cause of LBD is not fully understood, but it is linked to the accumulation of alpha-synuclein protein, with risk factors including age, male gender, and a family history of Parkinson's or dementia. Diagnosis involves neurological exams, imaging tests like MRI or Positron Emission Tomography (PET) scans, and assessments of cognitive functions, although no definitive cure exists. The treatment of LBD focuses on symptom management through cholinesterase inhibitors for cognitive decline, and physical and occupational therapies to maintain mobility.

Figure 1.6 shows microscopic view of brain tissue with a marked Lewy body, an abnormal protein aggregate linked to neurodegenerative diseases like LBD. Such formations disrupt neuronal function, contributing to cognitive and behavioral symptoms [21].

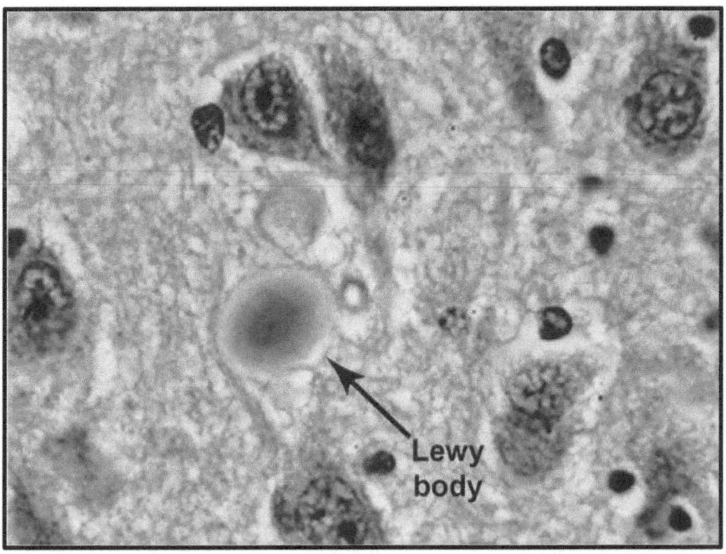

Figure 1.6: Lewy body in brain tissue (microscopic view)

1.2.6. Alzheimer's Disease (AD):

AD is recognized as a progressive neurodegenerative disorder, with cognitive functions of memory in particular, reasoning and spatial awareness being progressively impaired. It develops through steps, initially presenting as mild forgetfulness before progressing to severe cognitive dysfunction and loss of autonomy [6][22]. The symptoms of AD extend beyond memory deficits including confusion, impaired judgment and reduced problem-solving abilities, alongside personality changes and difficulty in recognizing close individuals.

As the disease progresses, challenges in performing routine tasks and maintaining coherent speech become increasingly obvious, ultimately resulting in complete dependency and dementia. The etiology of AD is multifactorial, with genetic predisposition, aging, and environmental influences identified as primary contributors [23].

Certain genetic mutations have been linked to increased susceptibility to AD, while lifestyle factors such as dietary habits, physical inactivity and chronic illnesses, particularly cardiovascular, have been implicated in the disease progression [6][22][24].

Effective detection of AD relies on advanced neuroimaging technologies. MRI identifies structural brain changes caused by neurodegeneration, while PET scan helps in detecting the presence of amyloid plaques [25].

1.3. Book Aims:

The main aim of this book is to classify AD into its various levels of ND, VMD, MD and MoD.

This study also aims to develop innovative hybrid methodologies to establish CNN architectures using advanced optimization techniques of Optuna and Keras-tuner. Such approaches facilitate the selection of optimal CNN architecture with well tuning its hyperparameters, leading to attain high performance.

Furthermore, a comprehensive medical system has been suggested and implemented. It utilizes efficient hardware and software components of RP5 and Python.

This system has the ability to provide advanced AD healthcare solutions as early diagnosing, offering risk assessment, enabling rapid intervention and affording immediate decision-making.

1.4. Book Contributions:

The book contributions can be highlighted as follows:

- Discovering best CNN architectures and hyperparameters for AD levels classification by employing the effective optimization techniques of Keras-tuner and Optuna.
- Using two approaches namely the Optimized CNN by Keras-tuner (OCNNKT) and Optimized CNN by Optuna (OCNNO) for diagnosing AD levels.
- Utilizing two datasets to confirm the robustness and successfulness of the used approaches. These are the Best Alzheimer Magnetic Resonance Imaging (BA-MRI) and Alzheimer's Disease Magnetic Resonance Imaging (AD-MRI).
- Suggesting and implementing an AD medical system that employes specialized hardware and software components, it has the capability to effectively classify and diagnose AD levels.

1.5. Book Organization:

The book is organized as follows:

Chapter 1: This chapter introduces the human brain, brain diseases overview, book aims, book contributions and book organization.

Chapter 2: This chapter provides the historical foundations and evolving perspectives, pathology, progression and impact, evolution of AD scientific understanding, reasons of AD and associated risk factors, symptoms of AD, diagnosing AD, progression of AD, and related work.

Chapter 3: This chapter illustrates the methodology, DL concepts, CNN concepts, optimization concepts and types (keras-tuner and optuna optimizations), used DL approaches, AD medical system and evaluation measurements.

Chapter 4: This chapter demonstrates the practical results and discussions, utilized datasets, implementations setup, performance evaluation of the used approaches, suggested medical system demonstrations and comparisons.

Chapter 5: This chapter declares the conclusions and future work.

CHAPTER TWO

GENERAL OVERVIEW ON ALZHEIMER'S DISEASE (AD)

2.1. Historical Foundations and Evolving Perspectives:

AD is globally recognized as health challenge, it affects millions of people. It was firstly identified by Alois Alzheimer in 1906, AD pathology is defined by amyloid plaques and neurofibrillary tangles. Emil Kraepelin formally classified AD in 1910 as a cause of early-onset dementia, with the definition later expanding to encompass late-onset cases, which is now the most common form. Throughout the 20th century, discussions focused on the role of brain lesions in differentiating various dementia subtypes, as Alzheimer, Kraepelin and Fischer, exploring the significance of neurotic plaques and neurofibrillary tangles.

By the mid-century, disagreements emerged regarding late-onset dementia classification, as British and American researchers debated whether it was an extension of AD or a distinct condition. Contemporary perspectives now emphasize a biological focus, identifying beta-amyloidosis as a hallmark feature [26].

Ongoing discussions contrast definitions from the Alzheimer's association, which prioritizes beta-amyloid pathology, and the International Working Group's clinic biological approach [26][27][28].

AD is currently the sixth leading cause of death in the world, with prevalence expected to rise significantly by 2050 [29]. The World Health Organization (WHO) predicts that the number of affected individuals will reach 114 million, placing substantial social and economic strain on healthcare systems [30]. In 2010, an estimation of 46.8 million people who are living with dementia will be associated with care costs of 818 billion dollars. By 2030, AD cases are expected to increase to 74.7 million, with costs potentially exceeding 2 trillion dollars [22].

AD progresses over decades, advancing through four distinct levels, with varying symptoms across AD levels which account for the majority of occurrences.

They are influenced by a combination of vascular health, environmental exposures and psychological stress. Current treatments focus on symptom management rather than halting neurodegeneration.

Preventative approaches highlight the reduction of cardiovascular risk factors, including hypertension, hypercholesterolemia, obesity and diabetes, as these conditions are strongly associated with both heart disease and dementia.

Given the complexity of AD, a single therapeutic intervention is unlikely to be suggested, with ongoing research aimed at slowing the disease progression and enhancing patient outcomes [28].

2.2. Pathology, Progression and Impact:

AD is a progressive neurodegenerative disorder and one of the leading causes of dementia among older adults [22]. It is marked by a gradual decline in memory, cognitive functions and daily functioning, starting with short-term memory loss, and progressing to severe impairments in behavior, language and social interaction.

The disease's progression results from a complex interplay of biological, environmental, and lifestyle factors. Key pathological features include the accumulation of beta-amyloid plaques, which disrupt neuronal communication, and tau protein tangles, which lead to neuronal damage and cell death [31]. In addition to these biological mechanisms, vascular, environmental and psychosocial factors play a critical role in shaping the risk of developing AD [32]. The disorder is classified into various levels ranging from mild cognitive impairment to advanced dementia, each with distinct clinical challenges. AD poses significant psychological and practical challenges, not only for patients but also for caregivers, who bear the burden of providing continuous care and support [33][34]. Although no definitive cure currently exists, ongoing research is focused on understanding disease mechanisms and exploring effective therapeutic interventions, with an emphasis on improving quality of life and delaying disease progression.

2.3. Evolution of AD Scientific Understanding:

The scientific understanding of AD has significantly evolved since its identification by Dr. Alois Alzheimer in 1906, shaping current insights into its pathology and clinical progression. Early research prioritized the study of brain alterations associated with the disease, such as amyloid plaques and neurofibrillary tangles, first observed by Dr. Alzheimer in the brain of the inaugural documented patient. These pioneering discoveries laid the groundwork for comprehending the condition [35].

With subsequent technological progression, medical imaging techniques, including MRI and PET, have significantly enhanced the accuracy of diseases' diagnoses and the ability to monitor their progression. These tools have enabled scientists to observe Alzheimer's-related brain changes with greater precision [36].

Recently, genetic research has been contributed in identifying genes associated with an elevated risk of developing AD, providing a deeper understanding of its hereditary factors and mechanisms.

It has advanced the understanding of the genetic factors contributing to the disease [23].

2.4. Reasons of AD and Associated Risk Factors:

Multiple reasons and associated risk factors have been identified as contributing to an increased susceptibility to AD.

Each of the primary reasons which involves in the development of this disease will be explained. Figure 2.1 illustrates the different reasons that cause AD [37].

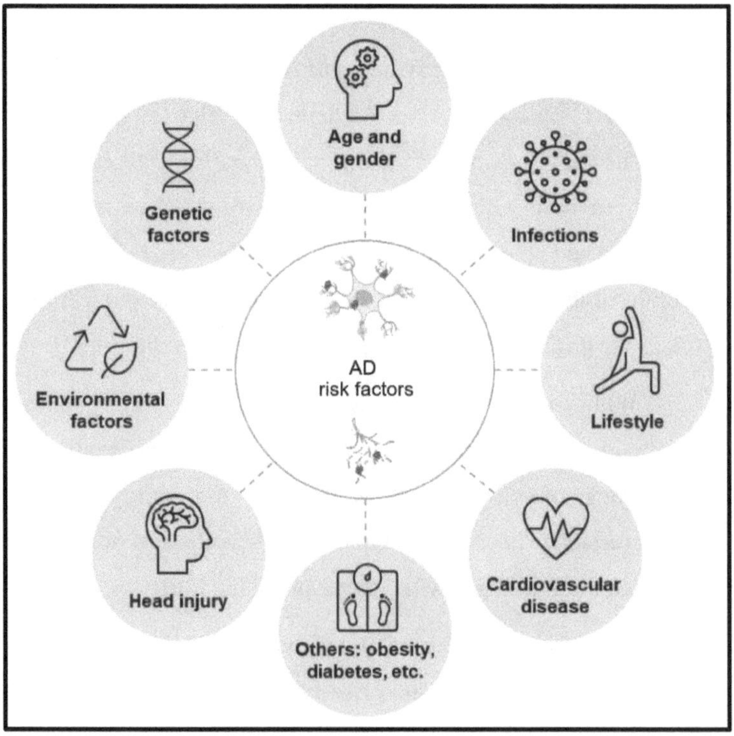

Figure 2.1: Different reasons of causing AD.

2.4.1. Genetic reason:

Genetic predisposition is considered as one of the most significant risk factors associated with AD. Several genes, including APOE-ε4, have been found to be linked to AD and identified as key contributors to its development [23][31]. The risk of developing the disease is believed to be higher in individuals carrying one or two copies of this gene than in those without that. Furthermore, rare familial cases of early-onset Alzheimer's have been linked to mutations in the PRESENILIN-1 (PSEN1) and PRESENILIN-2 (PSEN2) genes [38]. The genetic factors are thought to influence the formation and accumulation of beta-amyloid and tau proteins in the brain, leading to neuronal damage [39].

2.4.2. Environmental reason:

The role of environmental influences in the development of AD has been widely acknowledged. Prolonged exposure to toxic substances is believed to result in sustained neuronal damage. Additionally, long-term exposure to air pollution has been suggested to contribute to neuro-inflammation, thereby accelerating degenerative processes in the brain. It has been reported that lower levels of education and limited engagement in early life diminish cognitive reserve, leading to increase the brain's susceptibility to damage. Severe traumatic brain injuries have also been identified as potential environmental contributors to the disease [32].

2.4.3. Age and gender reasons:

Age is one of the most significant risk factors for AD. It has been observed that the condition commonly manifests after the age of 65, with its prevalence increasing steadily with advancing age [40]. Biological changes associated with ageing, such as increased inflammation and abnormal protein accumulation, significantly elevate the risk of AD. Women, particularly post-menopause, face a higher susceptibility due to reduced estrogen levels, which heighten neuronal vulnerability. Differences in symptom progression between genders highlight the need for tailored research and treatment approaches [40].

2.4.4. Lifestyle reason:

Lifestyle factors have been recognized as playing a considerable role in influencing the risk of AD [24]. Unhealthy dietary patterns, including high intake of saturated fats and sugars, alongside physical inactivity, have been linked to an increased risk of AD. Smoking and excessive alcohol consumption further contribute to neuroinflammation and disease progression [41].

2.4.5. Health issues reason:

Certain health conditions have been identified as major risk factors in the progression of AD. Chronic illnesses such as hypertension and arteriosclerosis are known to impair cerebral blood flow, leading to neuronal dysfunction and the accelerated accumulation of beta-amyloid and tau proteins. Increased inflammation and lipid imbalances associated with diabetes and obesity have also been linked to neurodegenerative processes. Furthermore, sleep disorders, depression, and chronic psychological stress have been found to contribute to cognitive decline by disrupting neuronal activity. Addressing these health concerns is considered essential for reducing the risk of AD and promoting brain health [42].

2.5. Symptoms of AD:

AD is defined by a spectrum of clinical symptoms and manifestations that develop progressively. these include memory loss, language impairments, behavioral and psychological changes, and difficulties with cognition and mobility. The severity of these symptoms varies among individuals and depends on the level of the disease. In the early levels, symptoms tend to be mild, such as occasional forgetfulness or challenges in finding the right words. However, as the condition progresses, these symptoms become more severe and begin to significantly interfere daily activities and quality of life [43][44]. Next, the most important symptoms of AD will be illustrated.

2.5.1. Memory loss symptom:

Memory loss is widely recognized as one of the earliest and most prominent symptoms of AD [45]. Initially, this manifest is as difficulty in recalling recent events, such as missed appointments or forgotten conversations, while older memories often remain intact for longer periods. Over time, individuals with Alzheimer's may struggle to remember essential information, including the names of close family members or basic daily tasks.

This decline in memory function is attributed to damage in the hippocampus and other brain regions responsible for information storage and retrieval [43][44][45].

2.5.2. Language impairments symptom:

Impairments in language are another key feature of AD. Patients may find it increasingly difficult to locate appropriate words during conversations, resulting in frequent pauses or the use of incorrect terms [46]. As the disease progresses, the ability to construct coherent sentences or comprehend spoken language may deteriorate further.

These linguistic challenges can severely impact communication, leading to frustration and social withdrawal [43][44][46].

2.5.3. Behavioral and psychological changes symptom:

AD is also associated with significant behavioral and psychological changes. Affected individuals may experience mood swings, irritability, anxiety or depression, often without an identifiable cause [24].

In some cases, more severe symptoms such as hallucinations or delusions may be developed. Such changes can present considerable challenges for caregivers [43][44], as they often necessitate additional emotional and practical support.

2.5.4. Cognitive and mobility difficulties symptom:

Cognitive and mobility issues become increasingly apparent as AD progresses [47]. Patients may struggle with spatial awareness, such as recognizing familiar places or judging distances.

These difficulties can make navigation increasingly hazardous, even within known environments. Consequently, individuals are at heightened risk of becoming lost or experiencing falls, posing additional threats to their overall well-being and safety [47].

2.6. Diagnosing AD:

The diagnosis of AD requires a comprehensive and methodical approach to accurately identify its condition and eliminate other potential causes of similar symptoms [48].

Achieving an early and precise diagnoses is pivotal, as it facilitates the formulation of an effective treatment plan and enhances the overall quality of life for the patient [48]. Next, a detailed explanation of AD diagnostic will be provided.

2.6.1. Medical history and clinical diagnoses:

The diagnostic process begins with an in-depth review of the patient's medical history, which includes their current symptoms, family history and he patient's health records. A thorough clinical examination is then conducted to evaluate cognitive and behavioral functions such as memory, concentration and problem-solving abilities [27].

2.6.2. Cognitive diagnoses:

Tests like the Mini-Mental State Examination (MMSE), and evaluations of memory and attention are employed to ascertain the impact of the disease on mental abilities [49]. These assessments provide insight into the specific areas of the brain involved in thinking and learning.

2.6.3. Brain imaging diagnoses:

Brain imaging techniques are essential for the early and accurately diagnoses of AD, providing valuable insights into the structural and functional abnormalities associated with its progression [22][50]. The most important types of these diagnostic techniques are:

- MRI: This scan detects brain atrophy and structural abnormalities commonly associated with AD.
- PET: This scan identifies the accumulation of amyloid protein within the brain, which is a hallmark indicator of the disease.

2.6.4. Lumbar puncture diagnoses:

Cerebrospinal fluid is analyzed for the presence of amyloid and tau proteins, which contribute to the formation of plaques and tangles observed in Alzheimer's patients [15][27][51].

2.6.5. Neuropsychological diagnoses:

Comprehensive cognitive assessments are performed to evaluate memory, language, attention and planning skills. These tests are crucial for tracking disease progression and understanding its effects on daily living [49].

2.6.6. Advanced imaging techniques diagnoses:

Advancements in imaging modalities have significantly enhanced AD diagnoses, allowing for a more comprehensive evaluation of both functional and structural brain alterations [50]. The following techniques contribute to a deeper understanding of disease progression:

- Functional Magnetic Resonance Imaging (fMRI): This technique scans the activities in specific brain regions and evaluates changes that correspond to the stages of the disease.
- CT: This technique scans and highlights structural changes in the brain, such as volume reduction or fluid build-up.

2.6.7. Behavioral and psychological diagnoses:

This aspect focuses on evaluating changes in personality and behavior, including symptoms of depression or anxiety, which may be indicative of the disease [44].

2.7. Progression of AD:

AD progresses gradually, with its impact on cognitive functions increasing over time. Initially, subtle neurological changes occur within the brain before any symptoms become apparent.

As the disease advances, memory, cognition and behavior are increasingly affected, eventually leading to the deterioration of essential functions [52].

The classification of AD levels facilitates an understanding of its progression, allowing healthcare professionals and caregivers to administer appropriate support at each phase [25][53].

Figure 2.2 shows examples of MRI brain scans that illustrates the various levels of dementia progression. Next, a detailed explanation of AD levels will be presented.

2.7.1. Normal level (ND):

At this level, no significant cognitive decline is observed and brain functions remain intact. However, minor neurobiological changes begin to develop at a molecular level, occurring without the manifestation of clinical symptoms of dementia [54][55].

2.7.2. Early level (VMD):

At this level, individuals experience very mild cognitive difficulties including occasional memory lapses, challenges in recalling recently acquired information and difficulty retrieving words during conversations. In this level, subtle changes in cognitive abilities become noticeable, reflecting early signs of impairment [54][55].

2.7.3. Moderate level (MD):

Memory deficits become more pronounced, leading to difficulties in recognizing familiar individuals, comprehending complex instructions and managing routine tasks such as financial planning. Emotional instability, mood fluctuations and reduced engagement in social activities are common, necessitating an increased level of assistance in daily life [55][54].

2.7.4. Advanced level (MoD):

This level is marked by extensive cognitive decline, leading to profound disorientation, impaired communication, and complete dependence on caregivers for fundamental needs such as eating, dressing and mobility. Neurological decline significantly affects physical health, increasing susceptibility to infections and other medical complications [54][55].

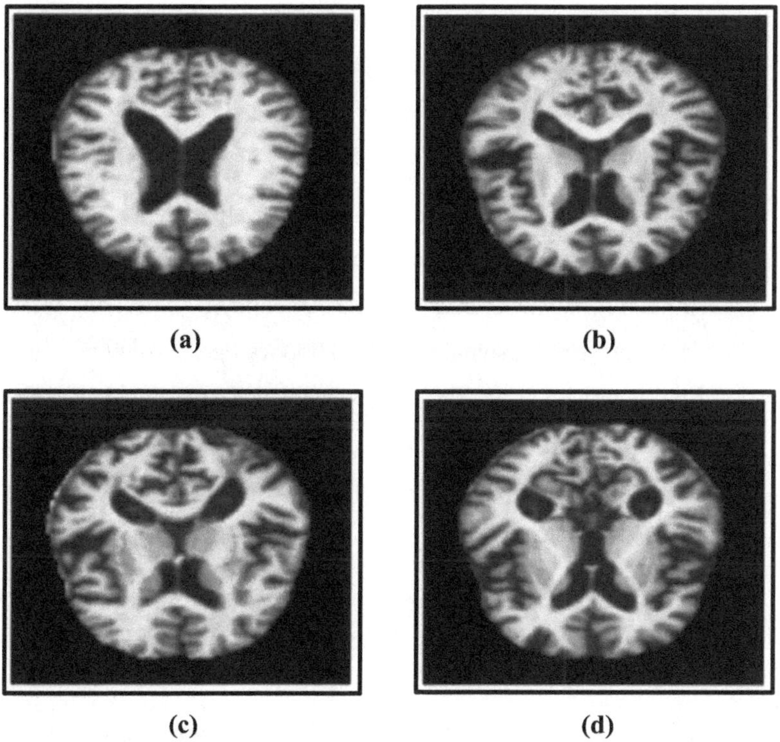

Figure 2.2: Examples of MRI scans for different levels of AD: (a) ND level, (b) VMD level, (c) MD level and (d) MoD level

2.8. Related Work:

A collection of previous studies that have explored the topic of AD can be presented as follows:

In 2013, Dyrba *et al.* demonstrated that machine learning algorithms enable accurate classification of Diffusion Tensor Imaging (DTI) data in spite of varying MRI devices. DTI metrics such as Fractional Anisotropy (FA) and mean diffusivity, along with gray and white matter density maps from anatomical MRI, were used in the classification process. Support Vector Machine (SVM) and Naïve Bayes (NB) algorithms were employed, and two cross-validation methods were applied namely pooled cross-validation and scanner-specific cross-validation. An accuracy of 80% was achieved for SVM with FA and 83% with MD. NB accuracies were ranged from 68% to 75% [56].

In the same year, Vos *et al.* conducted a study to predict Alzheimer for people with mild and moderate cognitive impairment. Biological and structural indicators, such as Beta-Amyloid (BA) protein, Total Tau (T-Tau) in cerebrospinal fluid, Hippocampal Volume (HCV) and Apolipoprotein E gene, were analyzed. The study included 399 subjects with mild cognitive impairment of dementia and 226 without. The results showed that the indicators significantly improved early diagnosis of AD [57].

In 2014, Sampath and Saradha highlighted the effectiveness of using an Adaptive Neuro Fuzzy Inference System (ANFIS) to classify AD levels.
K-means clustering was applied for segmentation and Gray Level Co-occurrence Matrix (GLCM) was used for feature extraction. MRI data for 150 subjects from the AD Neuroimaging Initiative (ADNI) dataset were analyzed. The subjects were categorized into Normal Controls (NC), Mild Cognitive Impairment (MCI) and AD. The results were compared between SVM and ANFIS classifiers. A higher accuracy of 88% was attained by ANFIS compared to the accuracy of 77.34% which was obtained by SVM [58].

In the same year, Fiscon *et al.* proposed a computational design for analyzing Electroencephalogram (EEG) signals to support diagnosing AD and MCI. Comprehensive analysis was identified to have broad applications, including basic analysis and time-frequency economic analysis. The dataset used in this study consisted of 100 samples. Various classification methods were employed such as SVM, decision tree and rule-based classifiers. The best accuracy of 90% was achieved by using the decision tree classifier [59].

In 2015, Demirhan *et al.* improved the accuracy of Alzheimer's classification by using feature selection techniques. Fractional anisotropy maps from the images were analyzed. Furthermore, averages in elliptical Region of Interest (ROI) and different voxel maps in elliptical fibers were used to drive SVM. The Relief Feature (ReliefF) algorithm was also employed to identify the most discriminative voxels in elliptical fibers. The total data used in this work was comprised of 221 samples. The work here increased the classification accuracy for all tasks by up to 15% [60].

In the same year, Gorji and Haddadnia introduced a novel method for the early diagnosis of AD based on Pseudo Zernike Moments (PZMs). Discriminative information was extracted from structural MRI images. Two types of artificial neural networks were employed: Pattern Recognition Network (PRN) and Learning Vector Quantization (LVQ) network. ADNI database of 500 images were utilized. The highest accuracy of 95.59% was achieved [61].

In 2016, Sarraf and Tofighi presented a method that utilized CNN to extract detailed features from brain images. Furthermore, LeNet-5 architecture was employed as an efficient clssifier. It was shown that the shift and scale-invariant features extracted by CNN followed by a DL classification such as LeNet-5 or Google Network (GoogleNet) could be exploited for more complex predictions. The database here, sourced from ADNI, included structural MRI images for 302 individuals over the age of 75. It comprised of 211 Alzheimer's patients and 91 non-Alzheimer's persons. was classified with an accuracy of 98.84% [62].

In the same year, Jha and Kwon developed a new method for diagnosing AD using Curvelet Transform (CvT) and K-Nearest Neighbors (K-NN) techniques. The CT was employed for extracting features from MRI images. Then, Principal Component Analysis (PCA) was applied to reduce dimensionality. After that K-NN along with Artificial Neural Network (ANN) were utilized for classification. MRI data from the Open-Source Imaging Studies Series (OASIS), including 416 individuals aged 18 to 96 years, was utilized. The proposed method achieved a classification accuracy of 89.47% [63].

In 2017, Hon and Khan suggested a strategy for the classification of AD using Transfer Learning (TL) with CNN architectures, including Visual Geometry Group of 16 (VGG16) layers and inception (version 4) networks. TL was utilized by leveraging pre-trained weights of the Image Network (ImageNet) database in order to reduce the need for a large training dataset. Image entropy was employed to identify the most informative slices of MRI data. OASIS database, which includes 6,400 MRI images from 200 individuals, was utilized. Classification accuracies of 92.3% and 96.25% were attained by VGG16 and inception, respectively [64].

In the same year, Gulhare and Sharma designed a technique for early detection of AD using a Deep Neural Network (DNN) architecture on brain MRI images. The images were segmented using the Niblack thresholding algorithm and features were extracted from the segmented regions using image processing techniques. A DNN classifier was then exploited to distinguish between normal tissue and AD tissue. The database here was for 150 individuals, 88 women and 62 men, aged between 60 and 96 years where 72 persons were without dementia and 78 persons were with dementia. The accuracy in this study was reached 96.6% [65].

In 2018, Islam and Zhang used Deep Convolutional Neural Network (DCNN) to diagnose AD by analyzing brain MRI data. Due to employing small data size, techniques such as data augmentation were utilized to improve the performance. Data from OASIS cohort were used, which included 416 MRIs of individuals aged 18 to 96 years, with focusing on patients who are older than 60 years. Different levels of AD were classified with the successful accuracy of 93.18% [66].

In the same year, Khvostikov *et al.* constructed a Three-Dimensional (3D) CNN to classify AD using structural MRI and DTI images. Advanced feature extraction techniques such as Gauss-Laguerre Harmonic Functions (GL-CHFs) were applied, which helped to identify ROIs in brain images. The Bag-of- Visual Words (BoVW) model was then used to encode the extracted features. A dataset of 280 samples was employed in this work. An accuracy of 92.5% was achieved in the results [67].

In 2019, Razavi *et al.* offered an intelligent AD diagnosis system using unsupervised feature learning. Sparser filtering was utilized to extract important features and reduced the influence of unhelpful features. Softmax regression was exploited for classification. The employed dataset of ADNI included 202 samples for the ages between 55 and 89 years, it involved 150 patients and 52 healthy individuals. An accuracy of 94% was attained here [68].

In the same year, Maqsood *et al.* designed an automated system for early detection and classification of AD using transfer learning techniques. The pre-trained convolutional network of AlexNet was utilized and the last layers were adapted to suit the classification of AD levels. The system was trained and tested on segmented and non-segmented images including Gray Matter (GM), White Matter (WM), and Cerebrospinal Fluid (CF) features. the system performance was evaluated using 382 3D MRI images of OASIS dataset, where a remarkable accuracy of 92.85% was obtained for the classification of non-segmented images [53].

In 2020, Mahajan *et al.* provided an overview of how machine learning techniques can improve the accuracy of AD diagnosis. SVMs were utilized to distinguish patient categories, while ANNs were employed to mimic brain processing. DL techniques like convolutional networks were reviewed for feature extraction from MRI and PET scans. Iterative-based feature selection techniques were used to ensure the classification efficiency. With this overview, the accuracy of diagnosis is significantly improved, verified, and validated [69].

In the same year, Rajak and Shrivastava investigated machine learning techniques to improve early detection and effectively diagnose AD. Such techniques were Logistic Regression (LR), SVM, Decision Tree (DT), Random Forest (RF) and Adaboost, which were aimed to be applied to MRI data and leveraged. The longitudinal OASIS dataset with 373 records and 15 features was used. The highest performances were achieved by the RF and Adaboost algorithm [70].

In 2021, Mahendran *et al.* focused on improving the classification of AD by using a hybrid pipeline of gene selection and DL. Minimum Redundancy Maximum Relevance (MRMR) strategy was used to reduce the redundancy between features and increase the relevance to target classes. Particle Swarm Optimization (PSO) was applied as a wrapper method to select relevant genes, with the Ki-Men algorithm to improve the results. An autoencoder was used to reduce dimensionality and compress data without significant information loss. Improved Deep Belief Network (IDBN) was employed to classify individuals for with and without AD. To enhance the accuracy optimal parameters for the IDBN model were determined using the Bayesian Optimization (BO) technique. The data were from the Gene Expression Omnibus (GEO) dataset, it included 161 samples. A high accuracy of 95.09% was reached [71].

In the same year, Bi *et al.* aimed to evaluate the performance and stability of a Deep Metric Learning (DML) algorithm in analyzing MRI images for AD diagnosis. An early diagnosis system was developed using both CNN and DML algorithms, with comparing their performances in classifying MRI images. Processing stages included image correction, bone removal, cross-entropy loss, and measurement loss. It was indicated that the DML algorithm could significantly contribute to the early diagnosis of AD, although increasing the sample size in future studies was suggested to strengthen the results. Data from the ADNI dataset were used, which included a total of 578 samples. 83% accuracy was achieved by this algorithm [72].

In 2022, Sethi *et al.* suggested a Computer-Aided Diagnosis (CAD) system for classifying AD using Two-Dimensional (2D) MRI slices. A hybrid model which combined CNN and SVM was proposed to boost the classification accuracy. MRI data were converted from Neuroimaging Informatics Technology Initiative (NIfTI) format into 2D slices in order to be used and analyzed. The middle slices from the extracted 2D images were chosen to ensure uniformity in the training and testing data. The ADNI dataset was exploited, it consisted of 1000 samples. An accuracy of 88% was attained here [73].

In the same year, Sheng *et al.* developed a technique for predicting AD using brain imaging and genetic data. MRI images were combined with genetic data to improve the performance. Fisher Score (FS) was utilized to reduce the dimensionality of genetic features before they were combined with imaging features. SVM algorithm was applied for classification and cross-validation was exploited to evaluate the overall model, it increased the reliability of the results. A notable accuracy rate of 98% has been secured with this technique [74].

In 2023, Mohi *et al.* designed a comprehensive CNN model to classify different levels of AD based on MRI data. Processing techniques such as rebalancing, data expansion, and preprocessing were applied to ensure the quality of the data used in training. CNN models such as Mobile Network (MobileNet), VGG16 and Residual Network of 50 (ResNet50) layers were used, and their performances were compared. Total of 2,900 MRI samples from ADNI were employed in this work The MobileNet model was improved, it obtained an accuracy of up to 96.6% [75].

In the same year, Khan *et al.* suggested transfer learning models of VGG16 and Visual Geometry Group of 19 (VGG19) layers to classify different levels of AD. A tissue segmentation technique was applied to extract GM features from MRI images. The GM was used to fine-tune the pre-trained VGG architecture with freezing features of the ImageNet database. This was accomplished by adding a new layer and gradually freezing some layers in the network. This work was done on 315 ADNI images and obtained notable accuracy [55].

In 2024, Chen *et al.* presented a classification technique based on the integration of multimodal data such as EEG, eye tracking and behavioral data to diagnose AD. Data augmentation method was implemented to enlarge the dataset. Extra Event-Related Potential Network (ERPNet) feature extraction layer was developed to extract multi-modal features. Domain Adversarial Neural Network (DANN) algorithm was utilized to improve the performance of diagnosis. Data from 69 participants were gathered at a hospital in China for this work. An average accuracy of 88.81% was reported [76].

In the same year, Bit *et al.* presented an algorithm of combining between Variational Autoencoder (VAE) technology and other machine learning classifiers to classify AD. VAE technology was used to extract low-dimensional featured vectors from MRI images. Machine learning models such as RF, extreme gradient boosting and SVM were exploited for the classification. Data from ADNI and OASIS datasets were utilized. An accuracy of 92% was obtained with this algorithm [77].

According to the surveyed prior studies and to the best of the obtained knowledge, no comprehensive work has been provided about exploiting an optimized CNN specifically for classifying AD levels. To address this gap, this study is presented. It uses two significant approaches, of the OCNNKT and OCNNO, which overcome the lacking.

CHAPTER THREE

METHODOLOGICAL CONCEPTS

3.1. DL Concepts:

3.1.1. DL principles:

DL is a branch of ML and AI, as shown in Figure 3.1 [78]. It is an advanced extension of ANNs. It processes large and heterogeneous datasets using specialized techniques and algorithms.

DL models can perform comprehensive classification directly from raw inputs [79][80]. Actually, DL provides significant difference to traditional ML.

Figure 3.2 demonstrates the conceptual difference between the workflows of traditional conventional ML models and DL [78][81].

That is, traditional ML workflow typically requires the consequences stages of: input, feature extraction, a traditional ML classifier and output. In this workflow, effective feature extraction is required for input data before moving to the ML classifier, such as: Support Vector Machine (SVM), Decision Tree (DT) or Logistic Regression Model (LRM).

In contrast, both feature extraction and classifier are integrated into a unified process in DL. That is, a DL model, such as CNN, can automatically extract and classify features from raw input data and provide the output. This end-to-end framework not only simplifies the required stages but also often leads to improved performances [78][79][81].

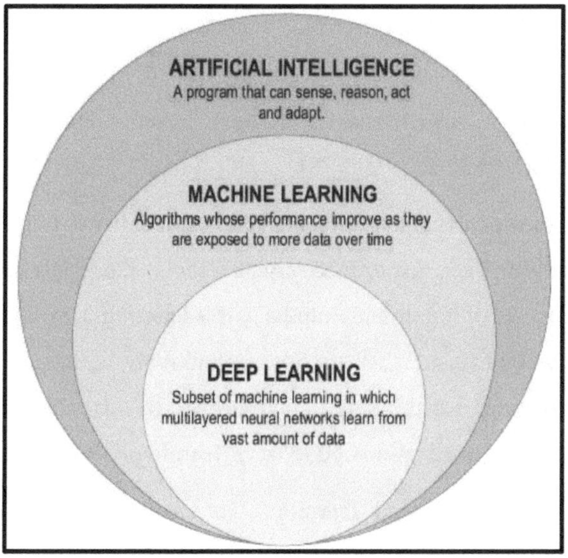

Figure 3.1: AI, ML, and DL: A Conceptual Overview of their Relationship

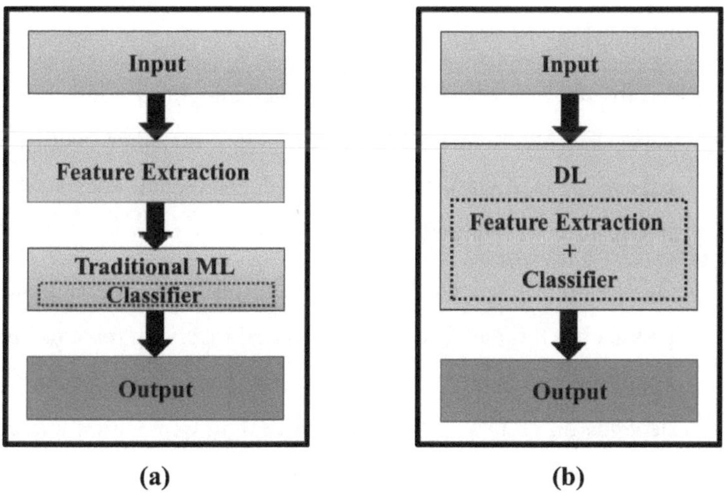

Figure 3.2: Conceptual difference between the workflows of traditional ML and DL (a) workflow with traditional ML and (b) end-to-end workflow with DL

3.1.2. DL types:

DL types can be categorized based on training into: supervised learning, unsupervised learning, semi-supervised learning and reinforcement learning [78].

In supervised learning, DL refers to a network that specifically handle data with their determined targets (labels). So, a DL network is provided with a set of inputs and outputs in this learning. During the training phase, the weights inside the network are iteratively adjusted to get more accurate estimations for the determined targets. In general, this strategy is more straightforward compared to other learnings [78][82].

In unsupervised learning, DL can be implemented without the presence of labels during the training phase. A DL network learns the important features or internal representations needed to discover connections between the input data. techniques like clustering are typically classified as part of the unsupervised learning paradigm [78][83].

In semi-supervised learning, DL integrates the aspects of both supervised and unsupervised learnings. In other words, the semi-supervised learning is based on semi-labeled data. Such strategy reduces the quantity of required labels [78].

In reinforcement learning, DL involves interacting with the application through a sequence of states, receiving rewards and selecting actions to maximize cumulative returns. The effectiveness of this process depends on factors such as reward design, exploration strategy and the complexity of the environment [78].

Basically, the aim of any DL type is to update the weights so that inputs are efficiently mapped to outputs. For each DL network, the training process adjusts its weights trying to reach reliable relationship between the input data and the required outputs. Subsequently, the final weights can be saved and later be applied to the testing phase, where they are used to generate predicted outputs for unseen inputs [78][82][83].

3.1.3. DL architectures:

DL architecture refers to the structural organization of a DL network, which are composed of many layers and neurons. Typically, such networks include multiple hidden layers between the input and output layers, enabling the autonomous extraction of features and the application of effective classifiers [78]. A prominent DL network example is the CNN, see Figure 3.3 [84].

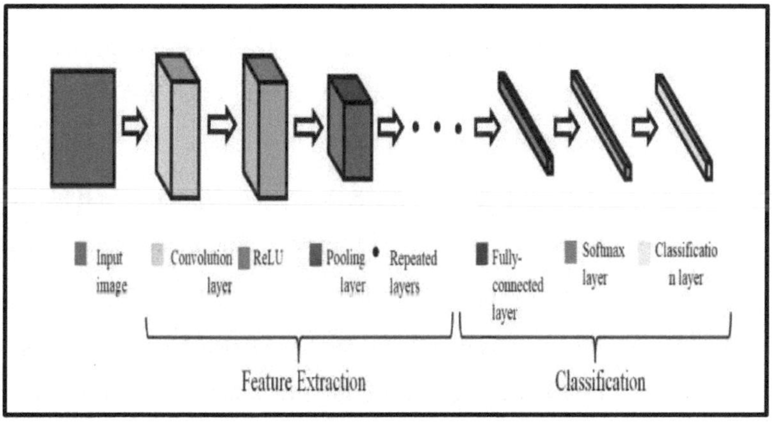

Figure 3.3: General architecture of CNN

CNNs are widely used in the tasks of image processing and recognitions. Their standard architecture consists of an input layer, series of hidden layers and output layer [84]. The hidden layers perform distinct functions such as convolution, Rectified Linear Unit (ReLU), pooling, dense or Fully Connected (FC), and softmax. These layers are generally grouped into two parts: a feature extraction part (including convolution, ReLU and pooling layers, often repeated) and a classification part (involving FC/FCs, softmax and classification layers) [78].

Other DL models based on CNN have been later established:

- XCM, YCM and ZCM networks: They are proposed and confirmed in recent publications. Each one of them composes of the principal CNN layers (input layer, convolution layer, ReLU layer, pooling layer, FC layer, softmax layer and classification layer). The differences between them represent by the specifications or determinations of their hyperparameters [85]. Main representative architecture for XCM, YCM or ZCM is given in Figure 3.4 [85].
- Visual Geometry Group of 16 layers (VGG16) and Visual Geometry Group of 19 layers (VGG19): Well-established models with 16 or 19 layers.

  Both include multiple convolutional, pooling and FC layers, and a softmax layer [86]. Their essential architectures are shown in Figure 3.5 [78][87].
- Residual Network of 50 layers (ResNet50): A deep residual network composed of 49 convolutional layers and a single FC layer [88]. ResNet50 general architecture is displayed in Figure 3.6 [78].
- Alex Network (AlexNet): A foundational CNN characterized by a relatively simple architecture consisting of 8 layers [89]. Main architecture of this network is given in Figure 3.7 [89].

- Densely Network of 121 layer (DenseNet121): A widely recognized DL network, it has an extensive architecture as it composes of 121 layers [78]. Its essential architecture is demonstrated in Figure 3.8 [78][89][90].

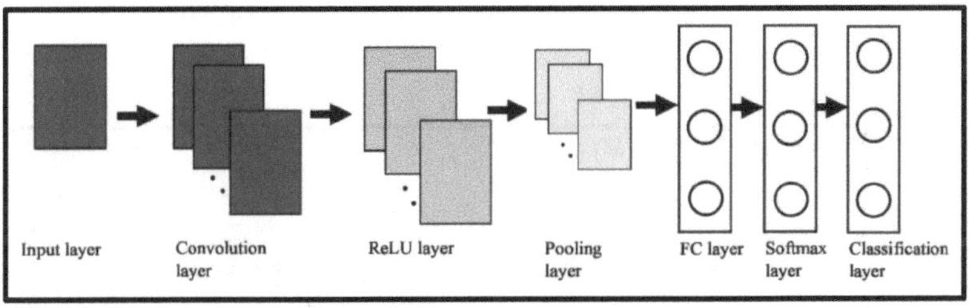

Figure 3.4: Main representative architecture for XCM, YCM or ZCM

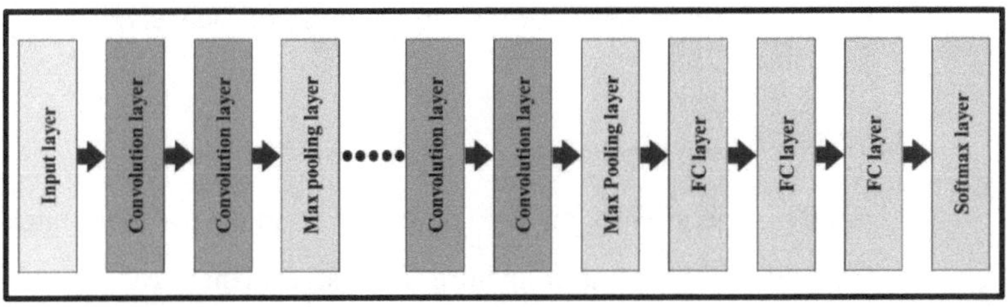

Figure 3.5: Essential architecture for VGG16 or VGG19

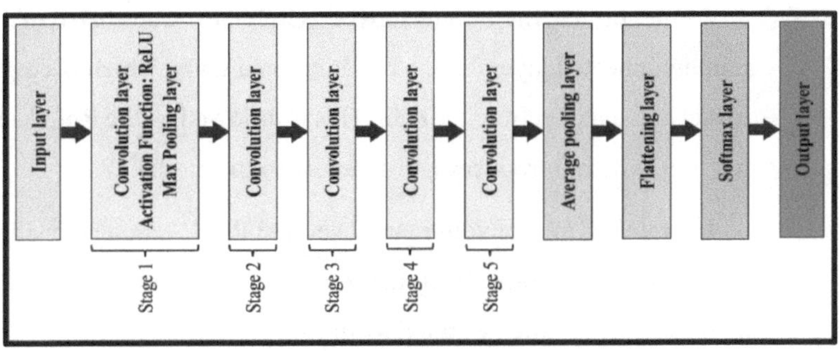

Figure 3.6: General architecture of ResNet50

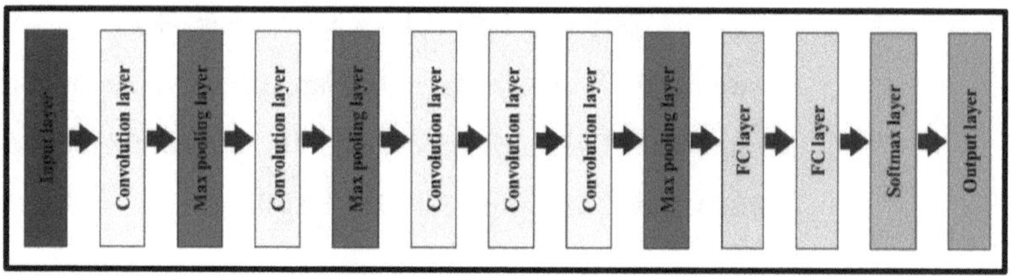

Figure 3.7: Main architecture of AlexNet

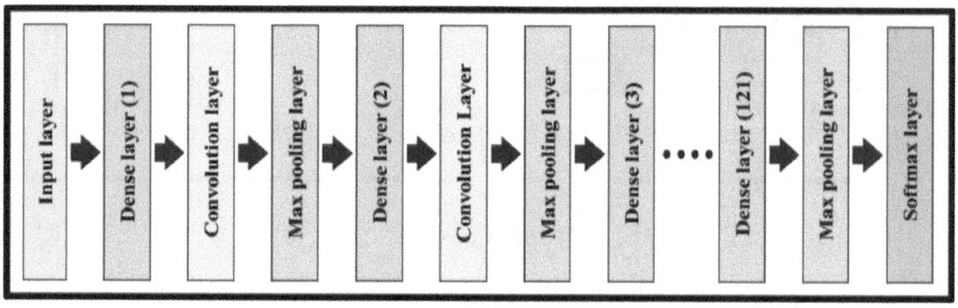

Figure 3.8: Essential architecture of DenseNet121

3.2. CNN Concepts:

CNN is specifically made to process data in the form of matrices, such as images [91]. Its architecture is distinguished by its exceptional capability to learn spatial patterns, making it exceptionally valuable at computer vision tasks, such as image classification and segmentation [92].

As mentioned, the architecture of a CNN can be consisted of a number of interconnected layers as the convolutional layer, which is employed to extract key features from the data. This is followed by the ReLU activation function, which introduces non-linearity to the model.

In practice, the convolutional layer and ReLU can be considered as a single layer. Subsequently, a max pooling layer is utilized to reduce the dimensionality of the extracted features.

The aforementioned layers can be repeated and are all used to extract features from images. The subsequent layers are responsible for classification. They may include a flattening layer, which is used to transform data into one-dimensional vector before feeding it into an FC layer (dense layer) [93]. The dense layer may use dropout ratio, which is utilized to prevent overfitting by randomly setting a fraction of input units to zero during training. Then, it passes the processed data to the final layer of softmax. In the softmax layer, final decisions are made based on the information received from the previous layer [78]. Next, is the detailed illustration of each layer.

(a) Convolutional layer: The convolutional layer plays a crucial role in extracting various features from an image. The convolution function can be represented by the following equation:

$$H_\alpha(a,b) = \sum_{m=1}^{M} \sum_{n=1}^{N} R(m,n) \times W(a-m\,,b-n) \tag{3.1}$$

where: $H(p,q)$ represents a calculated convolution value, M and N respectively represents the width and height of the Two-Dimensional (2D) input image, and $W(a-m\,,b-n)$ represents a kernel value [94]. It is important to note that the value of $H(p,q)$ is updated for a certain convolution (α) to be $H_\alpha(a,b)$, as numerous convolutional layers may be assembled within the proposed models. An example of convolution calculation is given in Figure 3.9 [78].

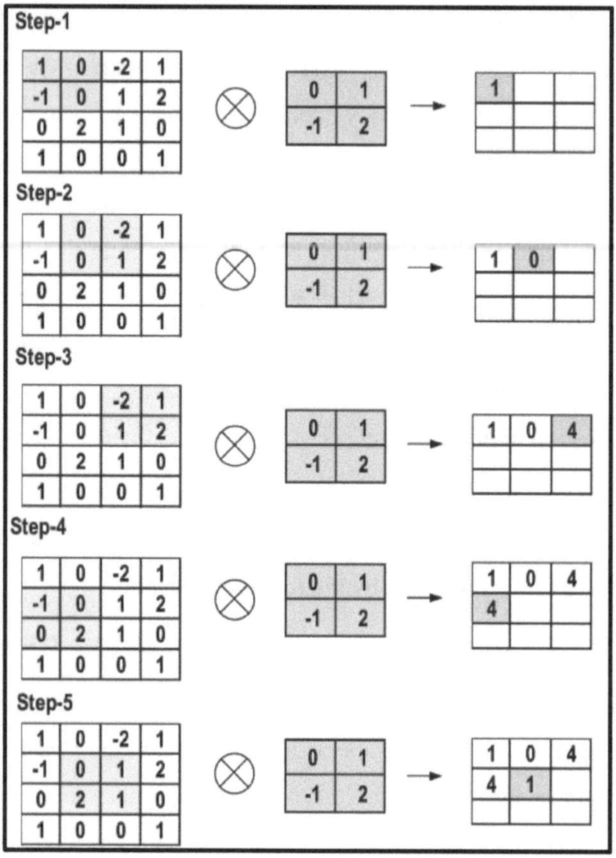

Figure 3.9: Example of convolution calculation with a kernel of 2×2 pixels, where the shape of ⊗ represents the convolution symbol

As mentioned before, the convolutional layer can be followed by ReLU, which is an activation function. The ReLU activation function is widely recognized for preserving positive values and eliminating negative values, thereby introducing non-linearity into the model.

This function can be mathematically expressed as follows:

$$Re_{\alpha}(x,y) = max(0, H_{\alpha}(a,b)) \tag{3.2}$$

where: $Re(x,y)$ represents a computed ReLU value and max represents the maximum operation [95]. Also, $Re(x,y)$ is updated for a certain ReLU (α) to be $Re_{\alpha}(x,y)$, as multiple ReLUs can be utilized in the multiple convolutional layers. Figure 3.10 shows an example of ReLU activation function implementation [78][95].

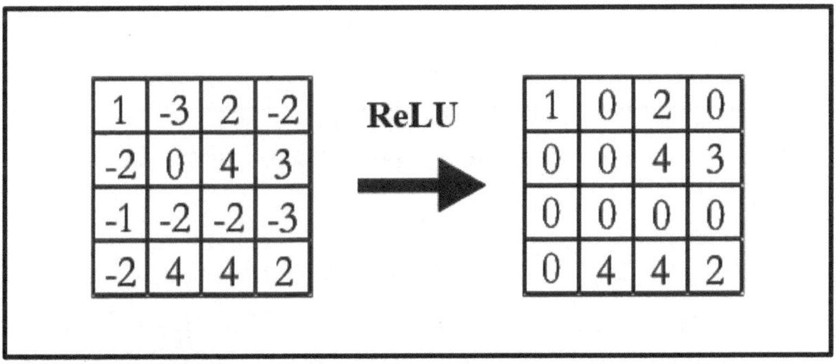

Figure 3.10: Demonstration of ReLU activation function implementation

(b) <u>Pooling layer:</u> The pooling layer reduces the dimensions of the previously analyzed vectors. This layer can be either of type max or average [78]. Max pooling is frequently utilized in CNNs. The general equation for the max pooling can be represented as follows [78]:

$$MP_{\beta}(k,l) = max(\mathbf{S}) \tag{3.3}$$

where: $MP(k,l)$ represents a computed pooling value and **S** is a small matrix (or a window) in a previous 2D vector [96]. Similarly, $P(k,l)$ is updated here for a certain pooling layer (β) to be $MP_{\beta}(k,l)$ as also multiple pooling layers can be established in the model. Figure 3.11 provides examples of applying each of max and average pooling [78].

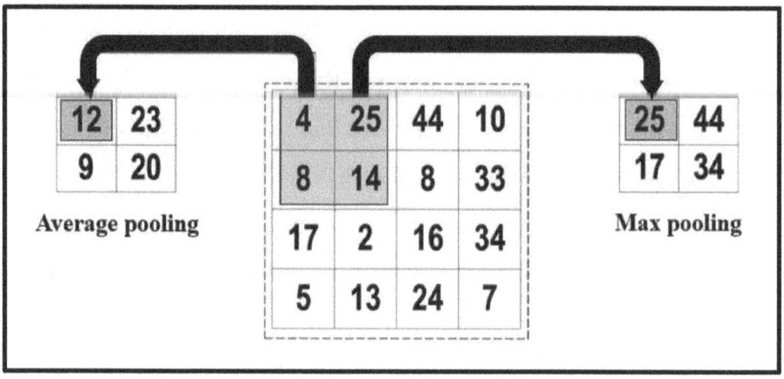

Figure 3.11: Examples of applying maximum and average pooling

(c) <u>Flatten layer:</u> The flatten layer converts a multi-dimensional vector into a One-Dimensional (1D) vector [78][93], preparing the data for the next layer (dense layer). The flatten layer equation can be represented as follows:

$$Fla[i] = \sum_{g=1}^{h} \sum_{f=1}^{w} \sum_{r=1}^{e} T[g][f][r] \quad (3.4)$$

where: $Fla[i]$ represents the flattened output for the ith sample, $T[g][f][r]$ refers to a value in a Three-Dimensional (3D) matrix, and h, w and e are the numbers of rows, columns and channels, respectively [93].

Flatten layer encapsulates extracted features of the previous layer, where they are transformed into 1D vector suitable for processing in the next layer [97]. Figure 3.12 demonstrates applying a flatten conversion [78][98].

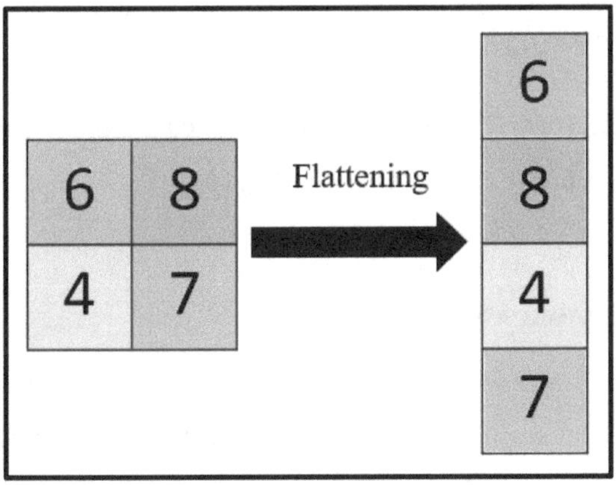

Figure 3.12: Demonstration of applying a flatten conversion

(d) Dense layer: The Dense layer (FC layer) connects each neuron in the previous layer to all neurons in its layer. The fundamental equation describing the fully connected layer can be expressed as follows:

$$DE_{\gamma}(c) = B(c) + \sum_{z=1}^{b}\left(FW(c,z) \times N(z)\right) \tag{3.5}$$

where: $DE(c)$ represents an output value of the dense layer, c represents the number of units (nodes) in the fully connected layer, $B(c)$ represents a bias value, b represents the number of units in the previous layer, $FW(c,z)$ represents a connection weight value between the pooling and fully connected layers, and $N(z)$ represents the output value of a node in the previous layer [78][99]. Likewise, $DE(c)$ is updated in this study for a certain pooling layer (γ) to be $DE_{\gamma}(c)$ as multiple dense layers can be provided in the model too.

(e) Softmax layer: The softmax layer can be considered as the output layer, it is ultimately employed for a specified classification task. Fundamentally, this layer utilises the softmax activation function. The core equation for this function can be expressed as follows:

$$SoMx(C) = \frac{e^{DE_{\gamma}(c)}}{\sum_{lo=1}^{K} e^{DE_{\gamma}(lo)}} \tag{3.6}$$

where: $SoMx(C)$ represents a calculated softmax value for class C and K is the number of classes [78][100][101].

3.3. Optimization Concepts:

Optimization is a fundamental discipline in applied mathematics and computer science that focuses on identifying the most effective solution to a given problem within a set of predefined constraints. It is widely employed across various fields including AI, economics, engineering, medicine, and logistics to either maximize desired outcomes, or minimize costs, errors or inefficiencies [102].

Any optimization model typically comprises of three core components [102]:

- Objective function: the function to be optimized, such as reducing error or maximizing profit.
- Variables: the parameters whose values can be adjusted to achieve the required solution.
- Constraints: the conditions or boundaries within which the variables must operate, such as resource limitations or time restrictions.

In the context of machine learning and artificial intelligence, optimization serves several critical roles [102][103]:

- Enhancing accuracy by reducing prediction or classification errors.

- Improving computational efficiency by minimizing memory usage and the number of required operations.
- Increasing speed by accelerating training and inference processes.
- Strengthening generalization, enabling models to perform effectively on unseen data.

This study exploits two types of optimization methods namely the Keras-tuner optimization and Optuna optimization, both of which have been applied in practice and proven their effectiveness in applying the aforementioned points.

3.3.1. Keras-tuner optimization:

Keras-tuner optimization was established as a modular and extensible system intended to facilitate the optimization of hyperparameters in DL models [104]. Its primary aim is to automate the search for optimal configuration of hyperparameters, thereby minimizing reliance on manual and computationally intensive tuning procedures [104].

Keras-tuner optimization has been employed to assist in the calibration and evaluation of a wide range of network configurations by systematically exploring the space of hyperparameters and assessing their empirical performance.

This process enables the identification of optimal parameter combinations that maximize model accuracy and stability within the scope of the study [105][104]. It enables tuning and evaluating a wide range of network configurations, aiming to identify the optimal parameters for achieving the best possible performance [104]. Figure 3.13 depicts the workflow of Keras-tuner optimization algorithm [104][105].

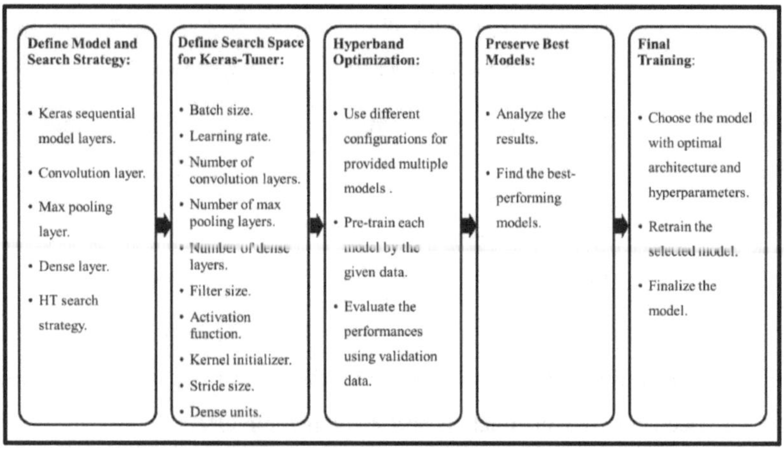

Figure 3.13: Block diagram of Keras-tuner optimization algorithm workflow

First of all, Keras-tuner algorithm utilizes one of four searching principals for hyperparameters optimization [105][106]:

- Random Search (RS): it samples configurations stochastically.
- Hyperband Technology (HT): it allocates computational resources dynamically using early stopping.
- Bayesian Optimization (BO): it leverages probabilistic models to guide search decisions.
- Custom Tuners Technology (CTT): it allows users to define tailored strategies.

Any of these searching principal enables scalable and efficient exploration of hyperparameters space in DL context, supporting both experimental flexibility and computational efficiency. In this study HT is used for the searching strategy [105][106][107].

HT can be utilized to speed up the process of identifying the model's architecture and its optimal hyperparameters. It is an advanced optimization technique designed to accelerate the hyperparameters tuning process for ML models [105]. By efficiently allocating computational resources, HT significantly reduces the time required to identify the optimal hyperparameters [105][106][107].

Keras-tuner optimization algorithm includes defining a network model that encompasses a series of tuning parameters [105][104][108]. These parameters include the batch size, learning rate, number of convolution layers, number of max pooling layers, number of dense layers, filter size, activation function, kernel initializer, stride size and dense units. They are known as the structural components of the neural network.

Keras-tuner initiates the process by generating multiple configurations of models for the predefined hyperparameters search space. The HT strategy is then employed to allocate computational resources efficiently after training each candidate for a limited number of epochs and evaluating them using validation accuracy [104]. That is, underperforming models in evaluations are discarded by HT, while it preserves the promising ones. This is repeated through trails. In each trial, a model is configurated and it is treated (trained with a limited number of epochs and evaluated) for multiple sets of parameters.

This strategy accelerates convergence toward obtaining optimal architecture and hyperparameters, with reducing redundant configurations. The model which achieves the highest validation accuracy is ultimately selected for deployment [107][108].

Upon the completion of the Keras-tuner optimization process, the final model is considered. It represents the optimal configuration in terms of structure and internal hyperparameters. Obviously, the exact structure of each model varies based on the used dataset.

3.3.2. Optuna optimization:

Optuna is an algorithmic optimization introduced in 2019 [109]. It is designed to enable dynamic constructions at runtime making it particularly effective for complex and condition-dependent models. It can enhance efficiency and reduce computations [109][110]. It incorporates a suite of robust sampling and evaluation strategies, including:

- Tree-structured Parzen Estimator (TPE): it is used for probabilistic inference [110].
- Covariance Matrix Adaptation-Evolution Strategy (CMA-ES): it is utilized for capturing and exploiting correlations between parameters.
- Random sampling: this is employed as a baseline approach or combined with custom optimization routines defined by the user.

In addition, Optima permits full customization of the optimization process, enabling users to incorporate their own strategies seamlessly [109][110][111].

Optuna optimization algorithm is basically designed to assist tuning DL networks [109][111]. It employs an intelligent technique to evaluate and fine-tune a wide range of parameters with the goal of identifying the optimal hyperparameters and the best possible architecture.

Optuna relies on the TPE for optimizing model's architecture and hyperparameters. TPE is a sophisticated technique that explores and incrementally refines parameters by learning from previous trials [110]. It employs a probabilistic optimization approach to enhance the efficiency of tuning. This method constructs a probabilistic model that distinguishes hyperparameters and configuration likely to yield high performance from those expected to perform poorly [109][110]. The method intelligently selects promising hyperparameters and architecture rather than relying on exhaustive or random search strategies [109][110]. As a result, it can reach high performance with a significantly reduced number of trials, making it especially useful in resource-constrained or time-sensitive applications [109][110][111]. Figure 3.14 displays the workflow of Optuna optimization algorithm.

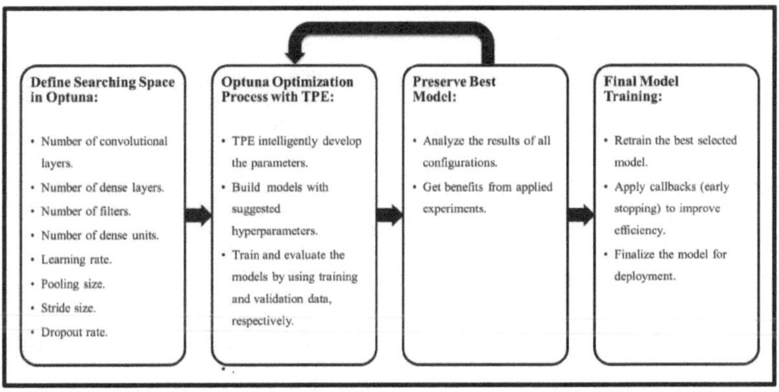

Figure 3.14 : Block diagram of Optuna optimization algorithm workflow

The algorithm begins by defining a network with key parameters [105][106][107][108]. These include the number of convolutional layers, number of dense layers, number of filters in the convolutional layers, number of units in the dense layers, learning rate, pooling size, convolutional stride size and dropout rate. Such parameters known as the structural components, they are optimized using Optuna, which works via trials [108][109] [111][112]. During each proceeding trial, the TPE technique is exploited to suggest new sets of hyperparameters and architectures based on the past trial [110].

Any established model is trained and evaluated based on the validation accuracy. A useful key idea here is that the Optuna has ability to develop a configured model in its algorithm by the power of the TPE [110]. This is implemented in each trial by getting experiences from previously applied experiments [110].

Such process repeats until the best-performing model, with the highest validation accuracy, is identified. Finally, the optimal hyperparameters and architecture for the final model are used [108].

Upon finalizing the optimization process, an integrated model is constructed, reflecting the best architecture and hyperparameters. This model is customized to deliver the optimal performance on the utilized dataset.

3.4. Used DL Approaches:

This study concentrates on integrating artificial intelligence methods for the detection and classification of AD by using brain MRI images. The proposed framework consists of a sequence of key stages.

The input stage involves supplying the system with a brain MRI image. This is followed by intelligent stage, in which one of the used approaches, OCNNKT and OCNNO, is employed to enhance the performance of the classification.

The final stage generates the output by assigning the MRI scan to a specific AD level or class, based on the model's prediction. Figure 3.15

shows the proposed framework block diagram of a used optimization approach.

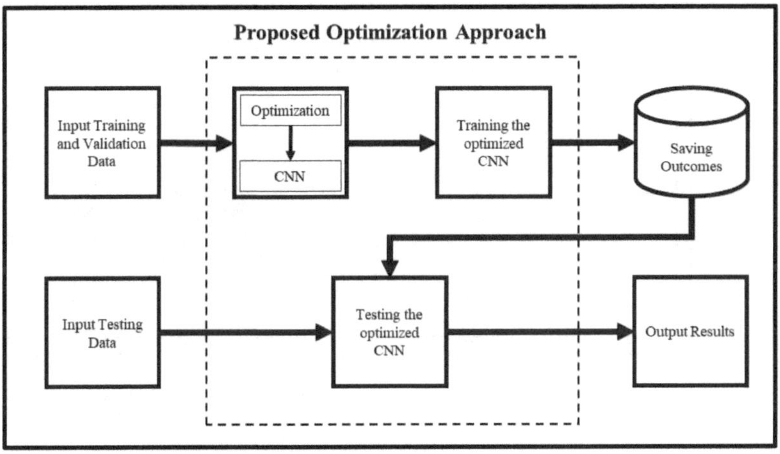

Figure 3.15: Proposed framework block diagram of a used optimization approach

3.4.1. Used OCNNKT approach:

In this book, the OCNNKT approach is used by integrating Keras-tuner optimization with CNN for the aim of diagnosing or classifying AD levels. The OCNNKT discovers a best CNN model by employing Keras-tuner optimization. In other words, Keras-tuner optimization is employed to select the optimal architecture of the CNN and efficiently tune its hyperparameters. This is implemented based on a specific goal that the optimization seeks which is to maximize the validation accuracy.

Figure 3.16 demonstrates the adopted steps with the used OCNNKT approach.

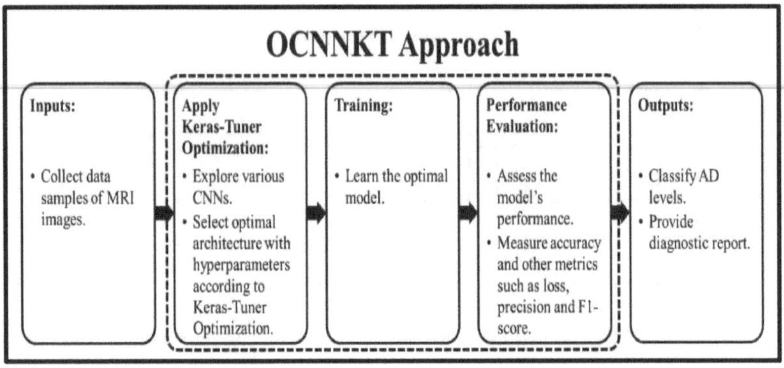

Figure 3.16: Adopted steps with the used OCNNKT approach

As two datasets of BA-MRI and AD-MRI are being utilized in this comprehensive study, two distinct models of CNNs are being constructed and employed by the OCNNKT approach. Each model is tailored to the specific characteristics of its respective dataset. Figure 3.17 shows the constructed architectures of the two models by the used OCNNKT approach.It can be noticed that the OCNNKT approach provides automated establishing of a CNN model which disregard the need for manual tuning to obtain such model. This also can enhance the model's efficiency, further reduce time consuming and enable efficient exploration for selecting the best CNN.

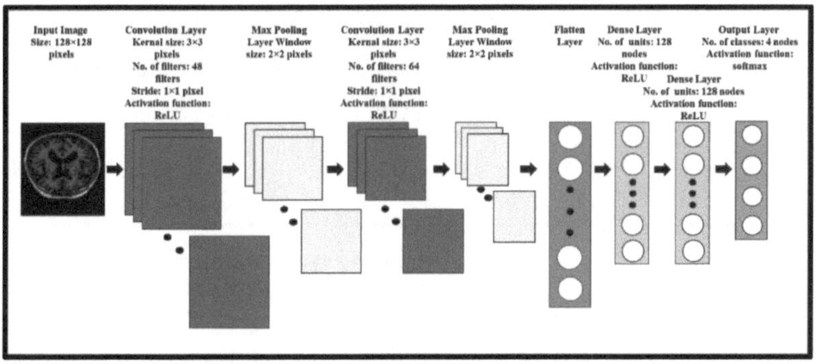

(a)

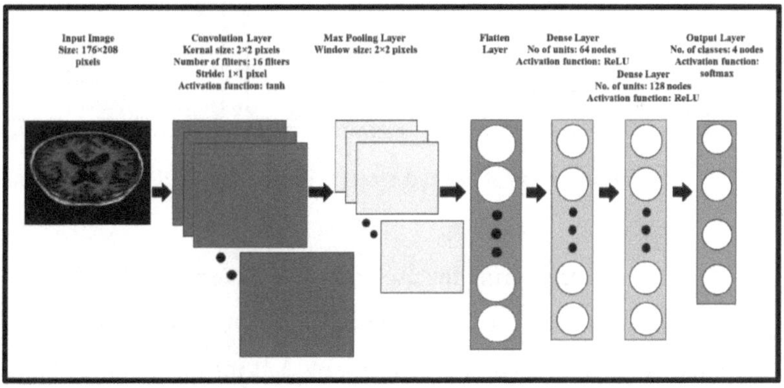

(b)

Figure 3.17: Constructed two models by the used OCNNKT approach, (a) for BA-MRI dataset and (b) for AD-MRI dataset

3.4.2. Used OCNNO approach:

In this study, the OCNNO approach is also used. It integrates or combines between Optuna optimization and CNN, aiming to diagnose and classify AD levels. Similarly, the Optuna optimization is exploited to select the optimal architecture of CNN and to efficiently fine-tune its hyperparameters. In other words, Optuna optimization is employed to judiciously select the best CNN architecture and proficiently fine-tuning the hyperparameters. Likewise, this is implemented based on discovering the goal of maximizing the validation accuracy. Figure 3.18 shows the adopted steps with the used OCNNO approach.

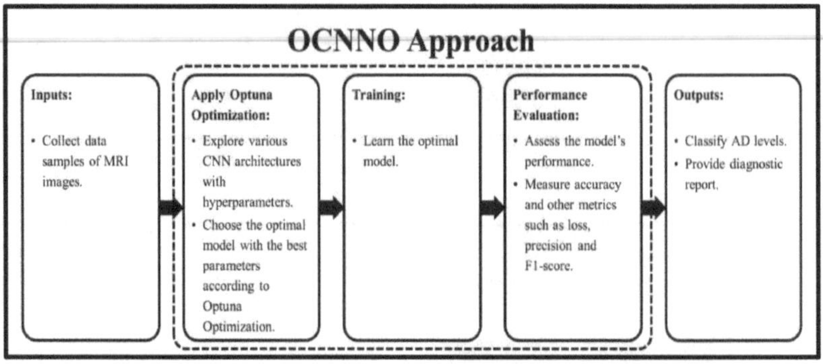

Figure 3.18: Adopted steps with the used OCNNO approach

Again, considering the use of the two datasets BA-MRI and AD-MRI, two distinct models of CNNs based on the OCNNO approach are obtained and employed. Each model is established for the specific characteristics of its respective dataset. Figure 3.19 demonstrates the constructed architectures of the two models by the used OCNNO approach. Reasonably, the OCNNO can further reduce the efforts and time consuming of looking for the best designed CNN. The automatic and dynamic capabilities of this approach allow the searching space to be defined at runtime, and its integrated pruning mechanisms halt unproductive trials early.

So, Optuna accelerates the construction of high performing models without requiring extensive manual intervention, resulting with the best establishment of network's architecture and hyperparameters.

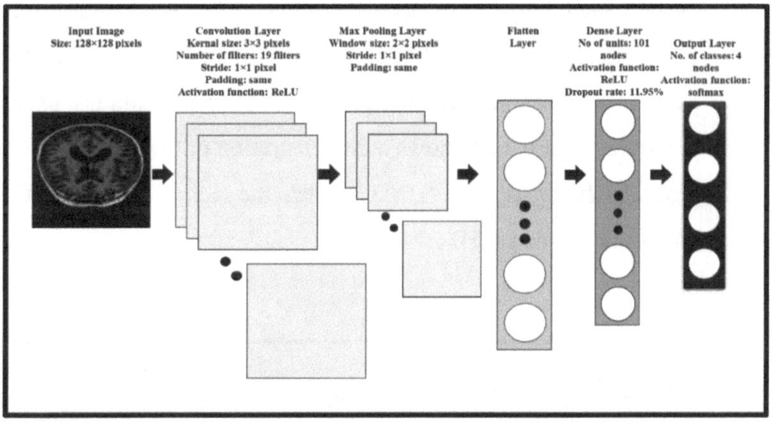

(a)

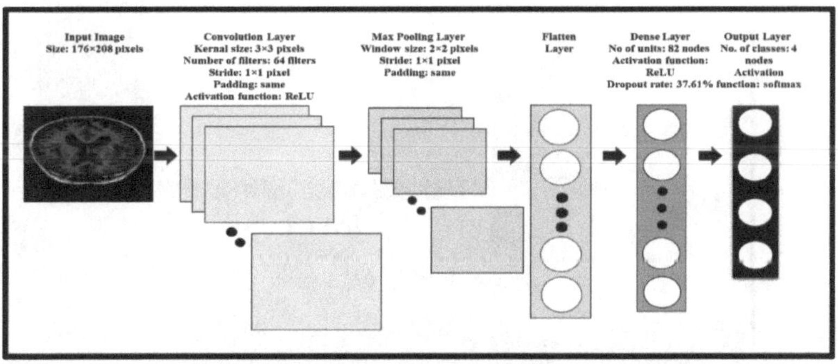

(b)

Figure 3.19: Constructed two models by the used OCNNO approach, (a) for BA-MRI dataset and (b) for AD-MRI dataset

3.5. AD Medical System:

A medical system is suggested for the diagnosis and classification of AD levels (ND, VMD, MD and MoD).

Figure 3.20 presents complete configuration of the suggested medical system, including all connected components. It has hardware and software parts. Raspberry Pi 5 (RP5) is an essential part of hardware and established Graphical User Interface (GUI) application is a useful part of software. Figure 3.21 shows a demonstration of the RP5 electronic board with clarifying its parts. Table 3.1 lists the technical specifications of its electronic board [113].

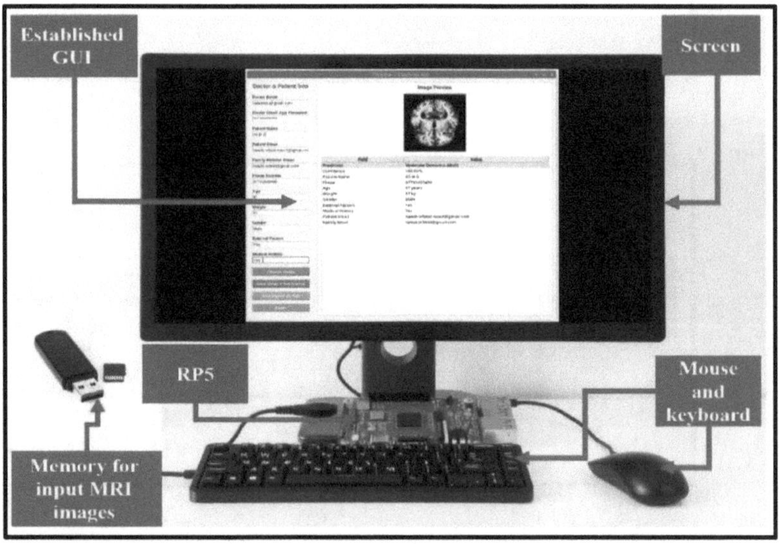

Figure 3.20: Complete configuration of the suggested medical system

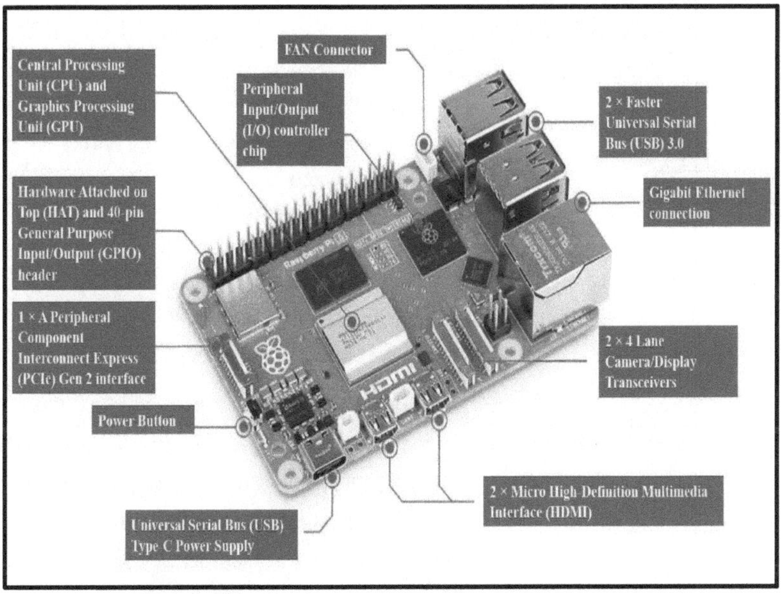

Figure 3.21: Demonstration of the RP5 electronic board with clarifying its parts

Table 3.1: Technical specifications of the RP5 electronic board

Specification	Value/Type
Electronic board	Raspbarry Pi 5 (RP5)
Processor	Quad-Core 64-bit ARM Cortex-A76 CPU clocked at up to 2.4 GHz
SDRAM	4 GB
Caches	o L1 Cache (32 kB Instruction Cache (per core)) + (32 kB Data Cache (per core)) o L2 Cache (512 kB Private Cache (per core)) o L3 Cache (2 MB Shared Cache (across all 4 cores))

Specification	Value/Type
Connectivity	o 2 × Micro-HDMI 2.0 ports (dual 4Kp60 HDR displays) o 2 × Universal Serial Bus (USB) 3.0 ports (5 Gbps each) o 2 × USB 2.0 ports o Gigabit Ethernet (1 Gb/s) with Power over Ethernet (PoE) + header support (via separate PoE + HAT)) o Dual-band 802.11ac Wi-Fi (2.4 GHz / 5 GHz) o Bluetooth 5.0 + BLE o 2 × 4-lane Mobile Industry Processor Interface(MIPI)Camera Serial Interface and Display Serial Interface (CSI/DSI) transceivers. o PCI-Express 2.0 x1 (onboard RP1 "southbridge") o 40-pin GPIO header (standard Raspberry Pi) o Dedicated fan header (JST connector) o Dedicated UART header (debugging / serial console) o Real-Time Clock (RTC) header (external battery-powered) o Composite video header (GPIO breakout) o MicroSD card slot (SDR104 high-speed mode) o USB-C power input (5V DC; up to 5A via PD support) o On-board power button

The system includes other hardware components. They can be illustrated as follows:

- A 42-inch display with High-Definition Multimedia Interface (HDMI) for visual interaction.
- Standard Universal Serial Buses (USBs) for keyboard and mouse, to be used as inputs for a user.
- A 65W power adapter for stable working.

Firstly, a brain image, acquired from an MRI machine, is input into the system through the established GUI software which can be operated using a standard keyboard and mouse.

This application is cross-platform, meaning it operates smoothly across various devices (computers, tablets and smartphones). Furthermore, the application uses the Flask web framework in Python programming language, which enables flexible and straightforward development.

The main window of the application prompts the user to enter basic patient information, including the patient's name, age, weight, gender, relevant external factors and medical history.

In addition, other useful information is also considered here, such as contacting details of the phone numbers and email addresses for the patient, trusted family member and attending doctor. Essential window of the established GUI application for the suggested medical system is given in Figure 3.22.

The system is facilitated with any optimized DL network of the used approaches, which is used to provide accurate and valuable AD diagnosis. When a brain image is uploaded to the established GUI application, the system automatically analyzes it using a pre-trained optimized model and then displays the diagnostic result. The result indicates whether the patient is affected by AD or not with identifying its specific level.

In addition, the system exploits the automated email notification, allowing to send alerts to the patient and trusted family member using the contact details entered in the main window.

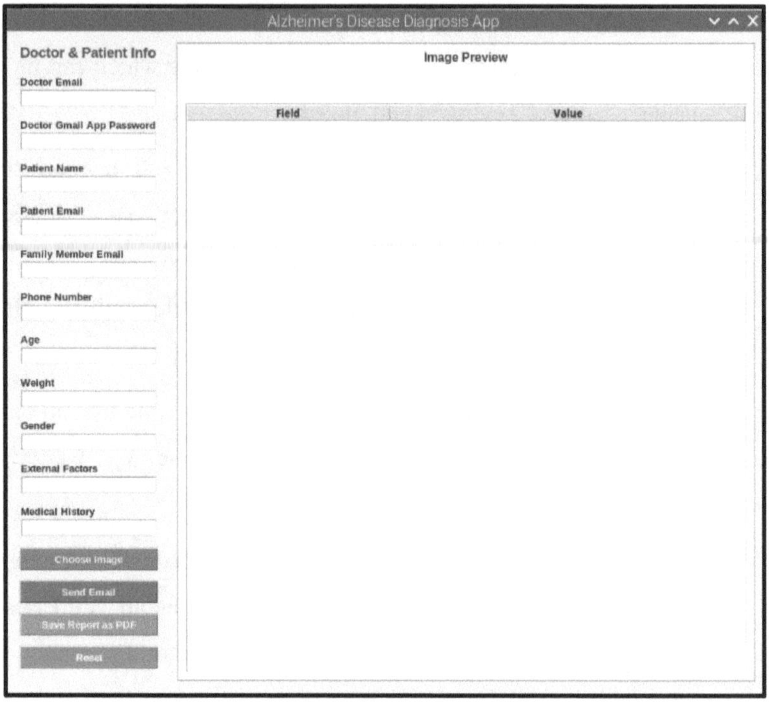

Figure 3.22: Essential window of the established GUI application for the suggested medical system

3.6. Evaluation Measurements:

The performance evaluations of the optimized CNN models, based on the used OCNNKT and OCNNO approaches, follow the same fundamental principles applied to other classification methods. The evaluation typically begins with the calculation of the four core components of the confusion matrix: True Positives (TP), True Negatives (TN), False Positives (FP), and False Negatives (FN). Specifically, TP denotes the number of positive instances correctly classified, TN represents the number of negative instances correctly identified, FP accounts for negative instances erroneously classified as positive, and FN denotes positive instances incorrectly labelled as negative [114][115][116]. These primary metrics serve as the basis for deriving more informative performance indicators.

One of the most fundamental evaluation metrics is accuracy, which reflects the proportion of correctly classified observations relative to the total number of predictions. It is computed as:

$$Accur = \frac{(TP+TN)}{(TP+TN+FP+FN)} \times 100\% \tag{3.7}$$

where $Accur$ represents the accuracy value, TP represents the TP, TN represents the TN, FP represents the FP and FN represents the FN [117][114]. Although widely used due to its intuitive nature, accuracy can be misleading in imbalanced datasets, where high values may not reflect the true performance of the model. To address this, more discriminative metrics are often employed.

Precision measures the proportion of true positive predictions among all instances predicted as positive. It is calculated as:

$$Prec = \frac{TP}{(TP+FP)} \times 100\% \tag{3.8}$$

where $Prec$ represents the precision value [118]. It provides insight into the classifier's ability to minimize false positives, which is particularly important in high-risk domains such as medical diagnosis.

Recall, also referred to as sensitivity, quantifies the model's capacity to correctly identify all relevant positive instances. It is expressed as:

$$Sens = \frac{TP}{(TP+FN)} \times 100\% \tag{3.9}$$

where $Sens$ represents the sensitivity value [118].

Conversely, specificity measures the proportion of correctly classified negative instances. It is considered as:

$$Spec = \frac{TN}{(TN+FP)} \times 100\% \quad (3.10)$$

where $Spec$ represents the specificity value [116].

Both sensitivity and specificity are critical in understanding the classifier's balance between detecting positive cases and avoiding false alarms.

A more comprehensive metric, f1-score, represents the harmonic mean of precision and recall. Thus, providing a balanced assessment of the model's performance, especially in scenarios with class imbalance. It is represented as:

$$F1\_sc = \frac{2 \times (Prec \times Sens)}{(Prec+Sens)} \times 100\% \quad (3.11)$$

where $F1\_sc$ represents the f1_score value [116].

Together, these metrics provide a multi-dimensional view of classification performance, enabling more nuanced assessments and comparisons between different applied methods.

CHAPTER FOUR

PRACTICAL RESULTS AND DISCUSSIONS

4.1. Utilized Datasets:

Two datasets are utilized in this study, they are called the BA-MRI [119] and AD-MRI [120]. Both datasets encompass of brain MRI images, which they are reviewed and examined by medical experts. Each image is of Joint Photographic Experts Group (JPG) format. The MRI images within each dataset are partitioned into the four AD levels of MD, MoD, ND and VMD. This categorization has been utilized for the classification purpose in this work. Table 4.1 provides detailed specifications for the BA-MRI and AD-MRI datasets.

Table 4.1: Detailed specifications for the AD-MRI and BA-MRI datasets

Specification	Type/Value for AD-MRI Dataset	Type/Value for BA-MRI Dataset
Source	Axial MRI scans	Original and synthetic axial MRI scans
MRI scanner strength	---	1.5 Tesla
Sequence	---	T1-weighted
Image resolution	208 × 176 pixels	128 × 128 pixels
Preprocessing	---	Skull removal
Total no. of images	2,609 images	11,519 images
No. of MD images	646 images	2,739 images
No. of MoD images	637 images	2,572 images
No. of ND images	652 images	3,200 images
No. of VMD images	674 images	3,008 images

BA-MRI dataset comprises of a combination of original and synthetic axial MRI scans. The scans were obtained using 1.5 Tesla MRI scanner with T1-weighted sequence. Each image has the resolution of 128×128 pixels. All images underwent preprocessing to remove skulls. In total, this dataset consists of 11,519 images. With breaking down: 2739 images for the MD, 2572 images for the MoD, 3200 images for the ND and 3008 images for the VMD [119].

AD-MRI dataset contains images captured using MRI scanner. Each is of the resolution 208 × 176 pixels. All images can be considered as raw data. In total, this dataset comprises of 2609 images. With breaking down: 646 images for the MD, 637 images for the MoD, 652 images for the ND and 674 images for the VMD [120].

4.2. Implementations Setup:

Firstly, all experiments have been conducted using a laptop with the following properties: type Dell, Intel core Trademark (TM) i5 processor of 8th generation which operates at the speed of 1.9 GHz, Randum Access Memory (RAM) of 16 GB and internal graphics card. Google Colab environment was also exploited for processing the two utilized datasets.

Initially, each dataset is divided into three distinct sets. A set for training, a set for validation and a set for testing with the proportions of 70%, 15% and 15%, respectively.

4.3. Performance Evaluation of the Used Approaches:

As mentioned before, two approaches are used in this study, which are the OCNNKT and OCNNO. Each approach has been analyzed for the four consecutive steps and phases of optimization, training, validation and testing. Detailed findings, visual illustrations and quantitative metrics are provided for both the AD-MRI and BA-MRI datasets.

4.3.1. OCNNKT Results and Discussions:

4.3.1.1 Optimization performances:

The OCNNKT dynamically optimizes of obtaining best CNN architecture and its hyperparameters, including the number of convolution layers, number of dense layers, filter sizes, strides and activation functions.

They are specifically tailored for a utilized dataset through multiple trails. Such trails are conducted autonomously by the OCNNKT, with the validation accuracy being evaluated in each trial. OCNNKT optimization results for both the AD-MRI and BA-MRI datasets are depicted in Figure 4.1.

Figure 4.1 presents the optimization results of the OCNNKT approach for the AD-MRI and BA-MRI datasets. It can be seen for the AD-MRI dataset that a steady improvement in the validation accuracy is observed, reaching around 90% after several trials which indicates effective convergence.

In contrast, limited progress in early trials for the BA-MRI dataset can be noticed in the validation accuracy but it is followed by a sharp improvement at the final trial, suggesting a more complex optimization landscape. These results confirm the adaptability of the OCNNKT approach in its optimization for different datasets.

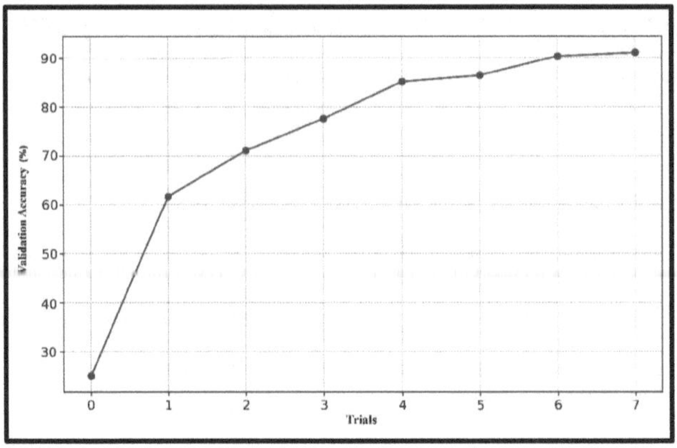

(a)

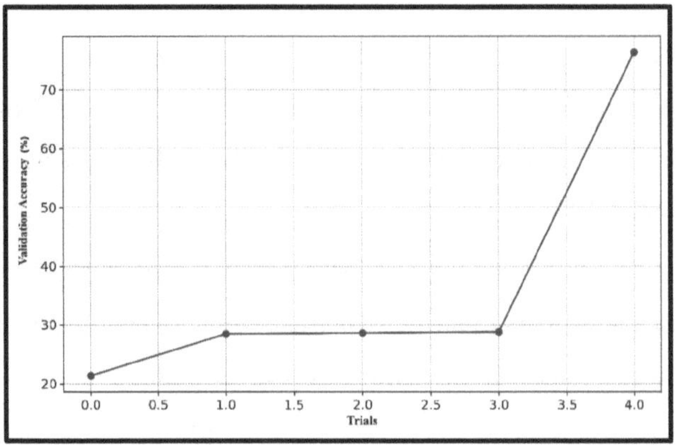

(b)

Figure 4.1: Efficacy of establishing OCNNKTs (a) for AD-MRI dataset and (b) for BA-MRI dataset

Upon the completion of the CNN model based on the OCNNKT, it is considered to be retained. The resulting configurations of the optimized CNNs for both datasets are given in Table 4.2 and Table 4.3. Both tables include all the determined parameters after using the Keras-Tuner optimization that contributes in constructing the optimal models of CNNs.

Table 4.2: Resulting configurations of the optimized CNN of OCNNKT for AD-MRI dataset

Layer Type	Kernal Size (pixels/nodes)	No. of Filters	Output Shape (pixels/nodes)	No. of Trainable Parameters
Convolution layer 1	**3 × 3**	**16**	**174 × 103 × 16**	**448**
Max pooling layer 1	**2 × 2**	**---**	**87 × 51 × 16**	**0**
Flatten layer	**---**	**---**	**70992**	**0**
Dense layer 1	**64**	**---**	**64**	**4543552**
Dense layer 2	**128**	**---**	**128**	**8320**
Dense layer 3	**64**	**---**	**64**	**516**
Total no. of parameters:				**4552836**
No. of trainable parameters:				**4552836**

Table 4.3: Resulting configurations of the optimized CNN of OCNNKT for BA-MRI dataset

Layer Type	Kernal Size (pixels/nodes)	No. of Filters	Output Shape (pixels/nodes)	No. of Trainable Parameters
Convolution Layer 1	**3 × 3**	**48**	**63 × 63 × 48**	**480**
Max Pooling Layer 1	**2 × 2**	**---**	**31 × 31 × 48**	**0**
Convolution Layer 2	**3 × 3**	**64**	**29 × 29 × 64**	**27712**
Max Pooling Layer 2	**2 × 2**	**---**	**14 × 14 × 64**	**0**
Flatten Layer	**---**	**---**	**12544**	**0**
Dense Layer 1	**128**	**---**	**128**	**1605760**
Dense Layer 2	**128**	**---**	**128**	**16512**
Dense Layer 3	**4**	**---**	**4**	**512**
Total params:				**1,650,980**
Trainable params:				**1,650,980**

4.3.1.2 Training phase performances:

During the training phase, the optimized CNN model tries to efficiently extract relevant features from the input data and simultaneously exploit its embedded classifier for the diagnosing task. These aim to minimize the disparity between predicted outputs and target labels.

Here, the optimized models are trained using the training sets. Very high training performances have been achieved for both utilized datasets, with the training accuracies of 97.47% for the AD-MRI and 98.94% for the BA-MRI. Figure 4.2 presents the training accuracy curves and Figure 4.3 depicts the training loss (error) curves.

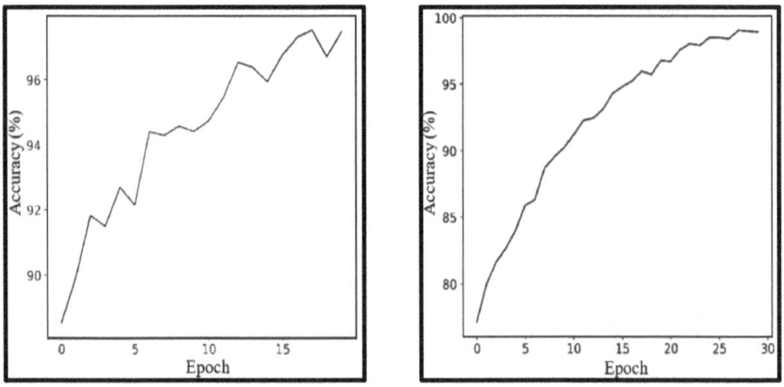

(a) **(b)**

Figure 4.2: Training accuracy curves of the optimized CNNs of OCNNKT (a) for AD-MRI dataset and (b) for BA-MRI dataset

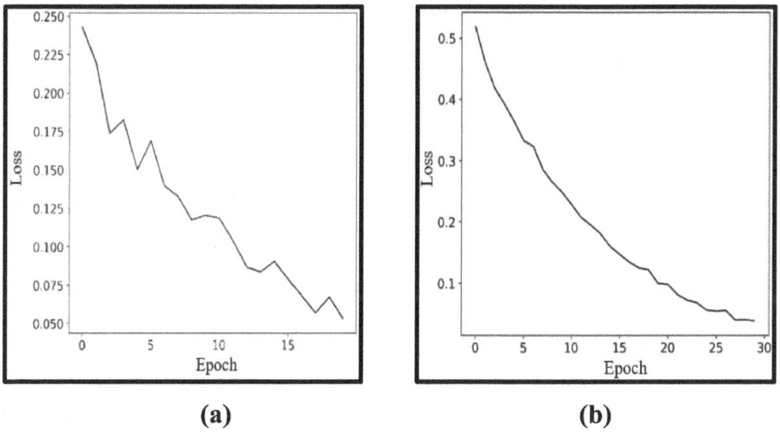

Figure 4.3: Training loss curves of the optimized CNNs of OCNNKT (a) for AD-MRI dataset and (b) for BA-MRI dataset

These figures show the training curves for the optimized CNNs based on the OCNNKT approach. Both models are demonstrated with rapid and acceptable improvements during their training progressions, indicating effective learning behaviors. The small numbers of epochs that are required to obtain the convergences reflect the efficiency of the used OCNNKT approach in selecting the highly appropriate models.

4.3.1.3 Validation phase performances:

The validation phase serves a critical role in evaluating a model's generalization ability using its set of data during the train. It can detect and prevent overfitting. Here, the trainings of optimized models are monitored by the validation sets to ensure validity. The results yield that high validation accuracies of 95.02% and 95.05% are recorded for the BA-MRI and AD-MRI datasets, respectively.

Such percentages reveal how well the OCNNKT approach implements for the validation stage.

4.3.1.4 Testing phase performances:

The testing phase represents the final stage of a model's evaluation, where the trained and validated model is assessed on a completely unseen dataset, referred to as the testing set. This phase provides an unbiased estimate of the model's generalization capability and real-time performance.

To evaluate any optimized CNN by the OCNNKT in the testing phase, the four core components of TP, TN, FP and FN are considered in every confusion matrix. So, determining the numbers of correctly and incorrectly classified AD levels in each class. Figure 4.4 shows the confusion matrices of the optimized CNNs for both employed datasets in the testing phase, illustrating the relationships between the actual (targets) and predicted (outputs) AD levels.

It is worth to highlight that the proposed models achieve excellent overall testing accuracies. Specifically, the overall testing accuracy of 92.44% is recorded for the AD-MRI dataset and the overall testing accuracy of 93.17% is benchmarked for the BA-MRI dataset.

Other statistical measurements are also considered for further evaluating performances of the optimized models. These measurements (or metrics) are: precision, recall and f1-score. Their obtained values are detailed in Table 4.4 and Table 4.5.

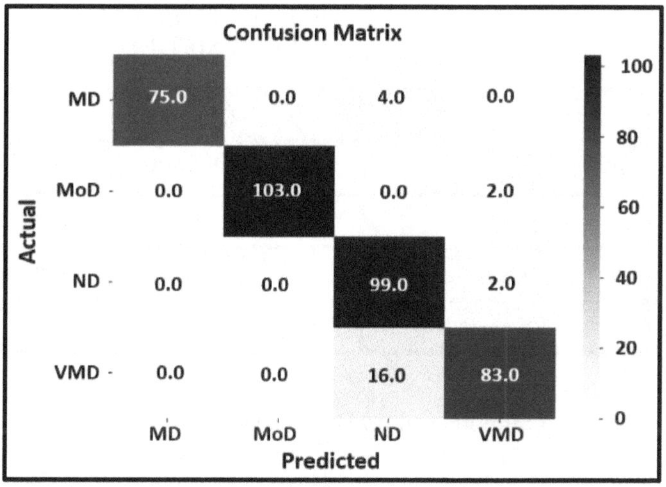

(a)

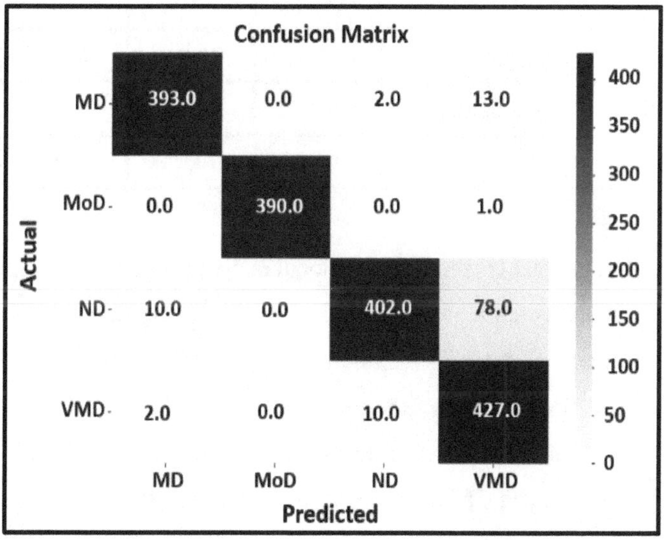

(b)

Figure 4.4: Confusion matrices of testing the optimized CNNs of OCNNKT (a) for AD-MRI dataset and (b) for BA-MRI dataset

Table 4.4: Statistical measurements of testing the optimized CNN of OCNNKT for AD-MRI dataset

Metrics	AD Levels			
	MD	MoD	ND	VMD
Precision	100%	100%	83%	95%
Recall	95%	98%	98%	84%
F1-score	97%	99%	90%	89%

Table 4.5: Statistical measurements of testing the optimized CNN of OCNNKT for BA-MRI dataset

Metrics	AD Levels			
	MD	MoD	ND	VMD
Precision	97%	100%	97%	82%
Recall	96%	100%	82%	97%
F1-score	97%	100%	89%	89%

These tables provide additional insight to evaluate the optimized CNN models which are constructed using the OCNNKT approach for the two utilized datasets. The results yield high values of precision, recall and f1-score, indicating the model's robustness in handling the classifications of AD levels. Figure 4.5 simulates mechanisms and performances of testing the optimized CNNs in classifying between various AD levels.

Such visual presentations emphasize the capabilities of the optimized models in diagnosing AD, where actual and predicted AD levels are matched.

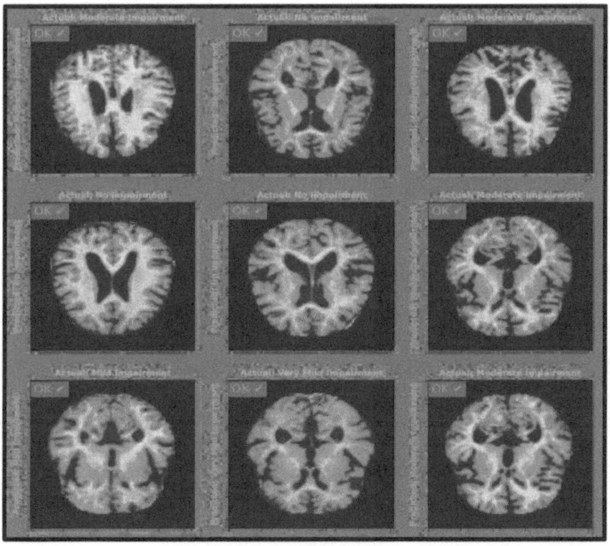

(a)

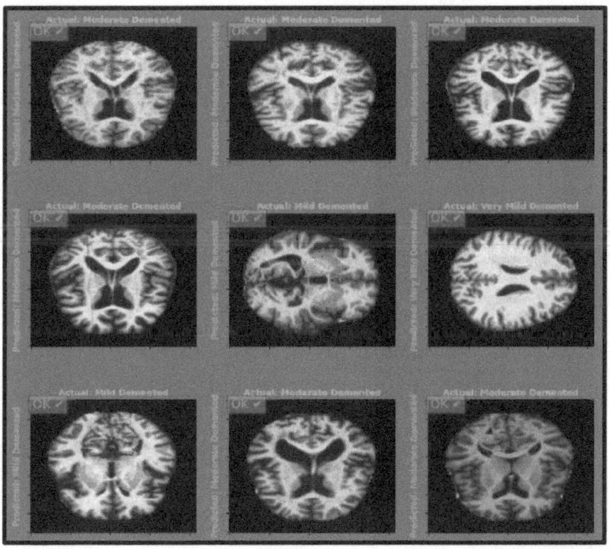

(b)

Figure 4.5: Simulating mechanisms and performances of testing the optimized CNNs of OCNNKT in classifying between various AD levels (a) for BA-MRI dataset and (b) for AD-MRI dataset

4.3.2. OCNNO Results and Discussions:

4.3.2.1 Optimization performances:

The OCNNO approach also dynamically optimizes of obtaining best CNN architecture and its hyperparameters, involving number of convolution layers, number of dense layers, number of filters, number of dense units, learning rate, pooling size, convolutional stride size and dropout rate. OCNNO performs its experiments of optimization with evaluating the validation accuracy in each trial.

OCNNO optimization results for both the AD-MRI and BA-MRI datasets are depicted in Figure 4.6.

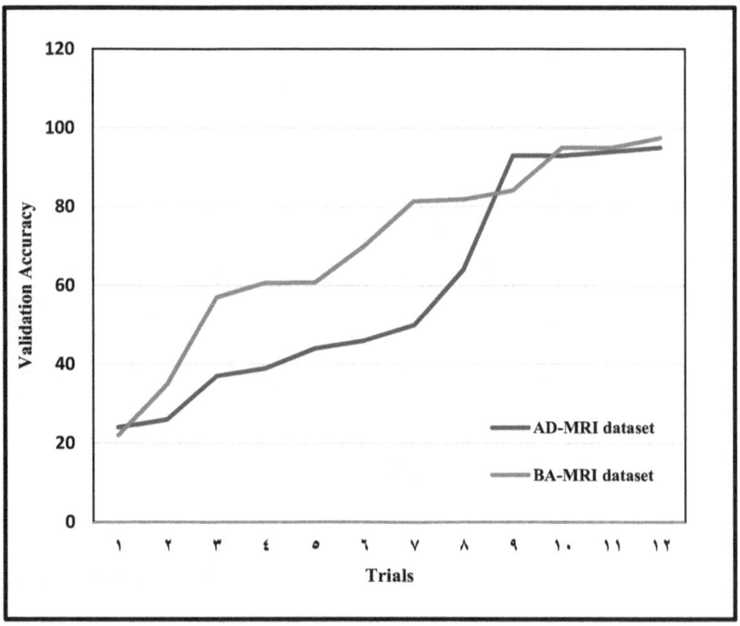

Figure 4.6: Effectiveness of validation during implementing the OCNNOs for both datasets

Then, the best optimized CNN model is considered. The resulting configurations of the optimized CNNs, based on the OCNNO approach, for both datasets are shown in Table 4.6 and Table 4.7.

Table 4.6: Resulting configurations of the optimized CNN of OCNNO for AD-MRI dataset

Layer Type	Kernal Size (pixels/nodes)	No. of Filters	Output Shape (pixels/nodes)	No. of Trainable Parameters
Convolution layer 1	**3×3**	**64**	**176 × 208 × 64**	**640**
Max pooling layer 1	**2×2**	**---**	**88 × 104 × 64**	**0**
Flatten layer	**---**	**---**	**585728**	**0**
Dense layer 1	**82**	**---**	**82**	**48029778**
Dropout layer 1	**82**	**---**	**82**	**0**
Dense layer 2	**4**	**---**	**4**	**332**
Total no. of parameters				**144092252**
No. of trainable parameters				**48030750**
No. of optimizer params				**96061502**

Both tables involve all the assigned parameters after utilizing the Optuna optimization which contributes in providing the best models of CNNs.

Table 4.7: Resulting configurations of the optimized CNN of OCNNO for BA-MRI dataset

Layer Type	Kernal Size (pixels/nodes)	No. of Filters	Output Shape (pixels/nodes)	No. of Trainable Parameters
Convolution Layer 1	**3 × 3**	**19**	**128 × 128 × 19**	**190**
Max Pooling Layer 1	**2 × 2**	**---**	**64 × 64 × 19**	**0**
Flatten Layer	**---**	**---**	**77824**	**0**
Dense Layer 1	**101**	**---**	**101**	**7860325**
Dropout layer 1	**101**	**---**	**101**	**0**
Dense Layer 2	**4**	**---**	**4**	**408**
Total no. of parameters				**23582771**
No. of trainable parameters				**7860923**
No. of optimizer params				**15721484**

4.3.2.2 Training phase performances:

In the training phase, the optimized CNN model of OCNNO also tries to efficiently extract important features from the input data and simultaneously utilize its classifier for the diagnosing of AD levels. Again, their goal is to minimize the differences between predicted outputs and targets. Logically, the optimized CNNs here are trained using the determined training sets. Remarkable outcomes and very high training accuracies are attained for the two utilized datasets. The training accuracies are reached 97.09% for the AD-MRI dataset and 99.47% for the BA-MRI dataset. Figure 4.7 presents the training accuracy curves for the two datasets and Figure 4.8 depicts the training loss (error) curves also for both datasets.

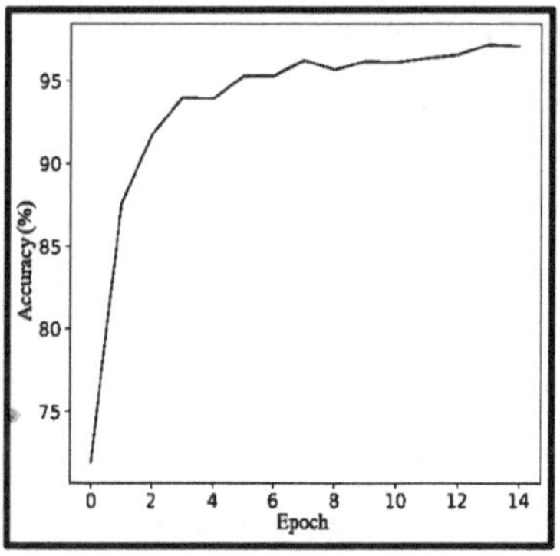

(a)

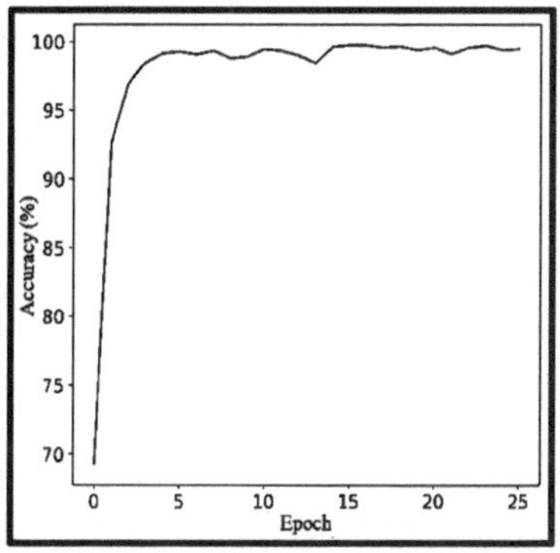

(b)

Figure 4.7: Training accuracy curves of the optimized CNNs of OCNNO (a) for AD-MRI dataset and (b) for BA-MRI dataset

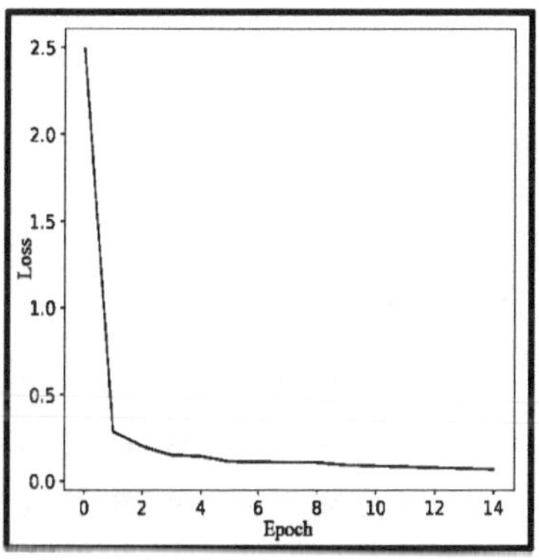

(a)

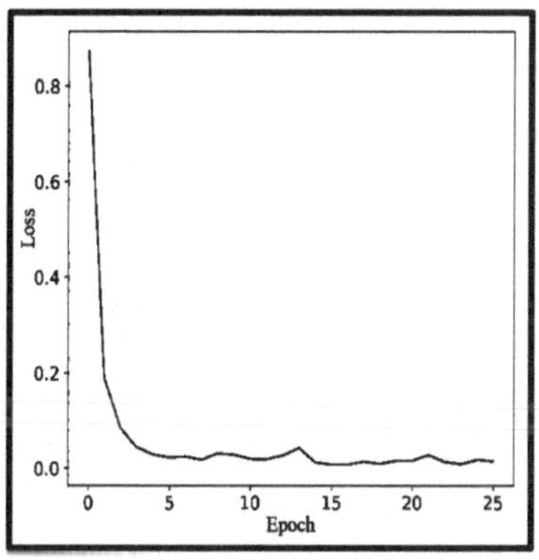

(b)

Figure 4.8: Training loss curves of the optimized CNNs of OCNNO (a) for AD-MRI dataset and (b) for BA-MRI dataset

These figures display the training curves of the optimized CNNs which are constructed by using the OCNNO approach. They demonstrate the rapid convergence and consistent improvements, reflecting efficient learning dynamics. Again, the small numbers of epochs required to accomplish the trains further highlights the effectiveness of the OCNNO approach in identifying well-suited configurations of CNNs.

4.3.2.3 Validation phase performances:

As mentioned, the validation phase considers evaluating a model's training capability by utilizing its set of data. It is valuable in preventing the overfitting. Here, the validation accuracies of the optimized CNNs, based on the OCNNO approach, after trainings are reached 99.53% for the BA-MRI dataset and 95.57% for the AD-MRI dataset. Such outcomes yield how well the OCNNO approach performs for the validation stage.

4.3.2.4 Testing phase performances:

As the testing phase represents the final and most critical stage in evaluating the effectiveness of the trained and validated models, it is considered for completely unseen dataset, known as the testing set.

This phase can offer unbiased estimation for a model to work under real-world conditions. To assess any optimized CNN by the OCNNO the in the testing phase, the four essential components of TP, TN, FP and FN are taken into account again for every confusion matrix. These metrics as such provide detailed breakdown of correctly and incorrectly classified AD levels for all classes.

Figure 4.9 shows the confusion matrices obtained during the testing phases of the optimized models for both the BA-MRI and AD-MRI datasets, they provide the relationships between the actual labels (targets) and the predicted outputs.

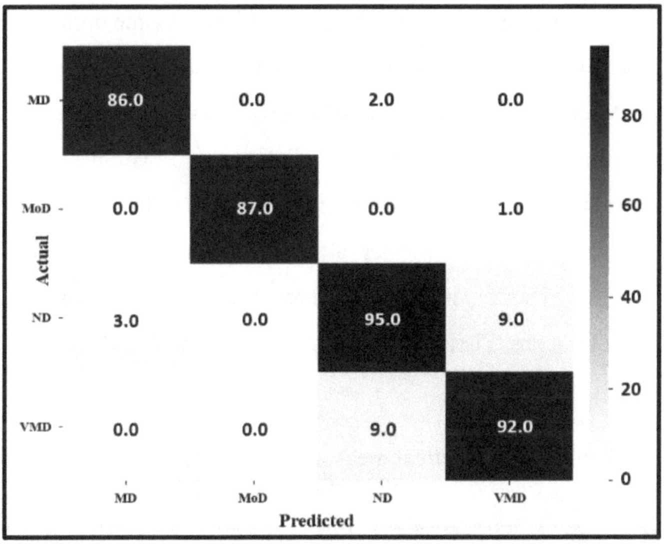

(a)

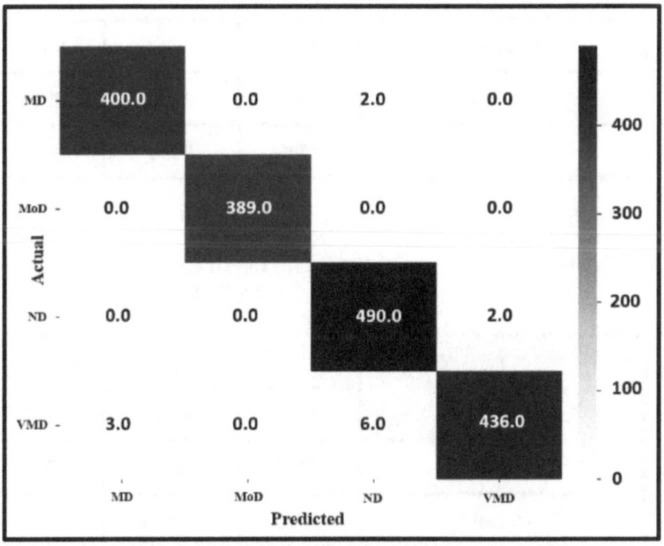

(b)

Figure 4.9: Confusion matrices of testing the OCNNO models (a) for AD-MRI dataset and (b) for BA-MRI datasets

The testing accuracies that are achieved by the optimized CNNs of the used OCNNO approach are benchmarked as 99.18% for the BA-MRI dataset and 95.79% for the AD-MRI dataset. Such results indicate strong classification performances and generalizations. Other statistical measurements are also taken into account for the used OCNNO approach, they can provide further assessments and insight for the optimized models. These statistical measurements (or metrics) are: precision, recall and f1-score. Their attained values are given in Table 4.8 and Table 4.9.

Table 4.8: Statistical measurements of testing the optimized CNN of OCNNO for AD-MRI dataset

Metrics	AD Levels			
	MD	MoD	ND	VMD
Precision	97%	100%	90%	90%
Recall	98%	99%	89%	91%
F1-score	97%	99%	89%	91%

Table 4.9: Statistical measurements of testing the optimized CNN of OCNNO for BA-MRI dataset

Metrics	AD Levels			
	MD	MoD	ND	VMD
Precision	97%	100%	97%	82%
Recall	96%	100%	82%	97%
F1-score	97%	100%	89%	89%

From these tables, it can be yield that the OCNNKT approach exhibits strong classification performances for both the AD-MRI and BA-MRI datasets as reflected by the high vales of precision, recall and f1-score. Consistent results are found for almost all classes of AD levels. Overall, the values of the additional metrics reveal reliable achievements by the optimized CNNs and their AD diagnostic capabilities, affirming their applicability in classifying the considered levels.

Figure 4.10 simulates mechanisms and performances of testing the optimized CNNs in diagnosing between the different AD levels. These visual presentations also highlight the abilities of the optimized models in classifying AD levels, where actual and predicted AD levels are corresponded.

4.4. Suggested Medical System Demonstrations:

As mentioned, a medical AD diagnosis system is suggested in this study. It integrates hardware and software components, such as RP5 microcontroller and Python language. RP5 is interfaced with GUI application and multiple peripheral devices, they are all valuable in using the system. The system is user-friendly and easy-to-use.

It can start with providing the GUI a brain MRI image and useful information such as the patient's name, age, weight, gender, relevant external factors and medical history. Contact details also can be offered such as phone numbers and email addresses for the patient, his/her trusted family member and attending doctor.

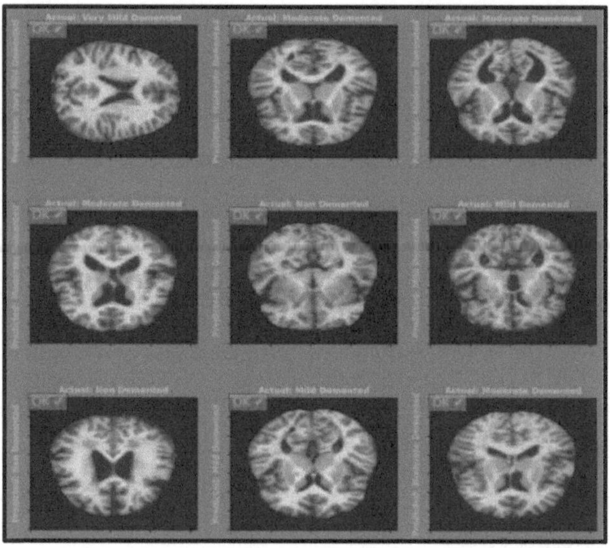

(a)

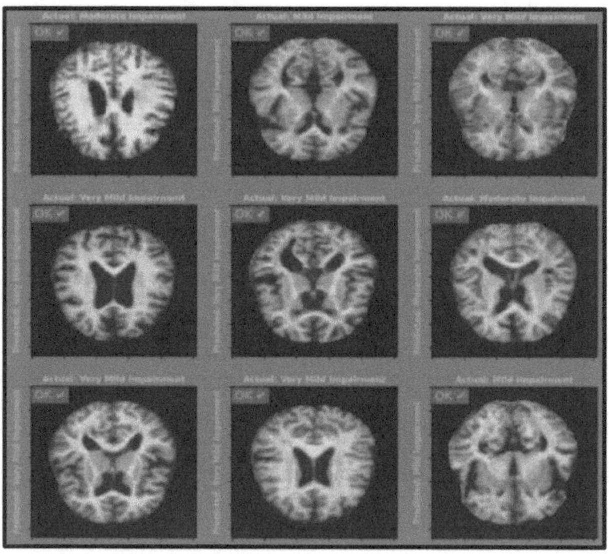

(b)

Figure 4.10: Simulating mechanisms and performances of testing the optimized CNNs of OCNNO in classifying between various AD levels (a) for BA-MRI dataset and (b) for AD-MRI dataset

The input images are subsequently processed during the DL stage, where each image is fed into the trained model to generate a diagnostic prediction. Additionally, the system supports automated email notifications directed to the patient, trusted family member and doctor, using the provided contact information at the beginning of starting the application. Furthermore, the suggested AD system has the facility of saving or exporting a medical report via the established GUI, it will be of type Portable Document Format (PDF).

Figure 4.11 shows the resulting window containing prediction outcomes, as well as, all the aforementioned information.

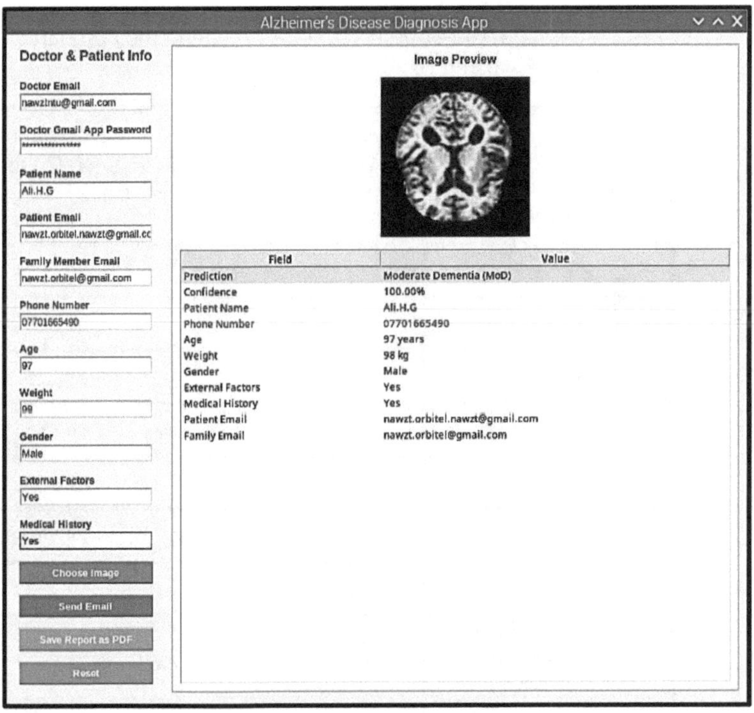

Figure 4.11: Resulting window containing prediction outcomes with useful information

4.5. Comparisons:

4.5.1. Comparison with other models:

Further investigations are conducted to compare between the performances of various DL models, including state-of-the-art, with the proposed optimized CNNs in the case of AD levels classification. For fair comparison, all models are trained, validated and tested with the same utilized datasets of the AD-MRI and BA-MRI for the same classification purpose. The compared DL models are: XAM, YCM, ZCM, VGG16, VGG19, ResNet50, AlexNet and DenseNet121. The classification accuracies of testing all considered models are summarized in Table 4.10.

Table 4.10: Classification accuracies of testing the compared models using AD-MRI and BA-MRI datasets

DL Models	Accuracies for AD-MRI Dataset	Accuracies for BA-MRI Dataset
AlexNet	71.15%	76.59%
ResNet50	75.55%	77.25%
VGG16	85.69%	88.29%
VGG19	87.69%	89.22%
DenseNet121	88.98%	90.15%
YCM	89.99%	88.55%
ZCM	90.11%	89.99%
XCM	90.12%	91.22%
Proposed optimized CNN based on OCNNKT	93.17%	92.44%
Proposed optimized CNN based on OCNNO	95.57%	99.53%

As it can be seen in this Table, some models as AlexNet and ResNet50 attain relatively poor accuracies for both datasets, such as 71.15% and 77.25%. This can be attributed to their excessive depth and layers, which are not necessary for the AD related data and diagnose.

In contrast, other models of VGG16, VGG19, DenseNet121, YCM, ZCM and XAM are composed of more valuable layers and parameters. Thus, it seems that they have better abilities to deal with diagnosing the AD levels. However, their results remain far from optimal, where their accuracies are ranging from 85.69% to 90.12% for the AD-MRI dataset and from 88.29% to 91.22% for the BA-MRI dataset.

Notably, the optimized CNN models surpass and outperform all the other compared models. As these models benchmark the highest classification accuracies for both datasets, confirming the efficiency and high capabilities of the used OCNNKT and OCNNO approaches.

4.5.2. Comparison between the used approaches:

Although the optimized CNNs based on both approaches of the OCNNKT and OCNNO attain the highest performances compared to other models, there are differences between their evaluations. That is, the models of the OCNNKT achieve the testing accuracies of 93.17% for the AD -MRI dataset and 92.44% for the BA-MRI dataset, whilst, the models of the OCNNO obtains further highest accuracies of 95.57% and 99.53% for the same datasets, respectively.

So, all optimized models based on the OCNNO benchmark better evaluations than all optimized models based on the OCNNKT.

This can be due the Optuna's ability to develop a configured model through its algorithm by taking into account getting benefits from applied experiments. Whereas, Keras-tuner has not used such ability through its algorithm. As a result, Optuna shows superior performances by providing more optimal models compared to Keras-tuner for the same utilized datasets and AD classification purpose.

CHAPTER FIVE

CONCLUSIONS AND FUTURE WORK

5.1. Conclusions:

This study has aimed to accurately classify AD into the levels of ND, VMD, MD and MoD. To achieve this, two approaches have been used for constructing the proposed CNN models, they are called the OCNNKT and OCNNO. These approaches have well optimized the architectures and hyperparameters of the CNNs, where advanced optimization techniques of Keras-tuner and Optuna have been employed. For confirming the robustness and successfulness of the work two datasets have been utilized, they are termed the AD-MRI and BA-MRI. Moreover, a medical system has been suggested for diagnosing the AD levels. The system has been designed with efficient hardware and software components, where RP5 and Python language have been exploited. The outcomes here can be concluded as follows:

- The proposed models of optimized CNNs based on the OCNNKT approach have obtained successful and consistent performances for both the AD-MRI and BA-MRI datasets. That is, the training accuracies have reached 97.47% and 98.94%, validation accuracies have attained 95.05% and 95.02%, and testing accuracies have achieved 92.44% and 93.17%, respectively.
- The proposed models of optimized CNNs based on the OCNNO approach have recorded high and remarkable performances for both the AD-MRI and BA-MRI datasets. That is, the training accuracies have attained 97.09% and 99.47%, validation accuracies have obtained 95.57% and 99.53%, and testing

accuracies have benchmarked 95.79% and 99.18%, respectively.

- Comparation with other DL models, including state-of-the-art, have shown that the proposed optimization models surpass all of them and for both datasets of the AD-MRI and BA-MRI. Eight DLs have been exploited for such comparison, these are the XAM, YCM, ZCM, VGG16, VGG19, ResNet50, AlexNet and DenseNet121.
- Additional comparison has been considered between the results of the OCNNKT and OCNNO approaches. It can be yield that all optimized models based on the OCNNO approach have outperformed all models of the OCNNKT approach. This superiority can be due to the Optuna algorithm capability in developing its model by getting experiences or benefits from applied experiments. In contrast, the Keras-tuner algorithm has no such consideration, instead it selects best models and discard others through its applied experiments.
- The AD medical system which has been suggested is useful and worth to be considered. It is friendly, easy-to-use and can demonstrate diagnosing the AD level with high reliability. It includes the effective GUI application which has been established with valuable facilities such as collecting helpful information (examples: medical history and relevant external factors), providing obtained outcomes, sending automated email notifications to the doctor and a trusted patient's family member, and exporting the medical PDF report.

5.2. Future Works:

Several directions for future work can be recommended to expand the volubility of this study, they are as follows:

1. Future studies may benefit from incorporating larger, more diverse MRI datasets drawn from multiple clinical centers.
2. The proposed framework can be adapted to detect and classify other neurodegenerative conditions such as Parkinson's disease, vascular dementia.
3. A remote-access AD diagnose system can be proposed, enabling clinicians to upload, analyze and interpret MRI images in real-time. Such facility may particularly benefit elderly patients or limited access individuals to specialists.
4. Assistance robotics can be connected to the suggested AD medical system to support patients in different daily life activities. For example, providing reminders and helps in taking medications, or in doing daily routines.
5. Incorporating other neuroimaging modalities and clinical metadata may improve diagnostic analyses and performances.

REFERENCES

[1] J. Hohwy and D. Harkness, "The Neural Organ Explains the Mind," *Open MIND, 2-vol. set*, vol. 19, 2022, doi: 10.7551/mitpress/10603.003.0058.

[2] K. P. Nair and K. P. Nair, "The Brain-Behavior Link: A Conundrum," *Food and Human Responses: A Holistic View*, pp. 25–44, 2020.

[3] Zalan Heszberger, Andras Gulyas, Andras Majdan, Andras Biro, Laszlo Balazs, Mezei Szabolcs and Jozsef Biro, "Proximity in the Brain," *Proceedings - 2020 International Conference on Computational Science and Computational Intelligence, CSCI 2020*, pp. 1378–1382, 2020, doi: 10.1109/CSCI51800.2020.00256.

[4] CVI Scotland, "Human Brain Parts." Accessed: Jun. 06, 2025. [Online]. Available: https://cviscotland.org/lessons.php?id=3

[5] Y. R. Lee, T. Q. Trung, B. U. Hwang, and N. E. Lee, "A flexible artificial intrinsic-synaptic tactile sensory organ," *Nature Communications*, vol. 11, no. 1, pp. 1–11, 2020, doi: 10.1038/s41467-020-16606-w.

[6] Samantha McGirr, Courtney Venegas and Arun Swaminathan, "Alzheimer ' s Disease : A Brief Review," *Journal of Experimental Neurology*, vol. 1, no. 3, pp. 89–98, 2020.

[7] Valery L Feigin and Theo Vos, "Global, regional, and national burden of neurological disorders, 1990–2016: a systematic analysis for the Global Burden of Disease Study 2016," *The Lancet Neurology*, vol. 18, no. 5, pp. 459–480, 2019, doi: 10.1016/S1474-4422(18)30499-X.

[8] Jin Liu, Min Li, Yi Pan, Wei Lan, Ruiqing Zheng, Fang-Xiang Wu, Jianxin Wang, "Complex Brain Network Analysis and Its Applications to Brain Disorders: A Survey," *Complexity*, vol. 2017, 2017, doi: 10.1155/2017/8362741.

[9] X. Wang and X. Huang, "Risk factors and predictive indicators of rupture in cerebral aneurysms," *Frontiers in Physiology*, vol. 15, pp. 1–18, 2024, doi: 10.3389/fphys.2024.1454016.

[10] Buoy Health, "Vascular Dementia." Accessed: Jun. 09, 2025. [Online]. Available: https://www.buoyhealth.com/learn/vascular-dementia#definition

[11] J. Adamson, S. E. Humphries, J. R. Ostergaard, B. Voldby, P. Richards, and J. T. Powell, "Are cerebral aneurysms atherosclerotic?," *Stroke*, vol. 25, no. 5, pp. 963–966, 1994, doi: 10.1161/01.STR.25.5.963.

[12] Saad Latifa, Muhammad Jahangeer, Dure Maknoon Razia, Mehvish Ashiq, Abdul Ghaffar, Muhammad Akramc, Aicha El Allamd, Abdelhakim Bouyahya, Larisa Garipova, Mohammad Ali Shariati, Muthu Thiruvengadamf and Mohammad Azam Ansari, "Dopamine in Parkinson's disease," *Clinica Chimica Acta*, vol. 522, pp. 114–126, 2021, doi: 10.1016/j.cca.2021.08.009.

[13] Shalby Hospitals, "Parkinson's Disease." Accessed: Jun. 16, 2025. [Online]. Available: https://www.shalby.org/parkinsons-disease/

[14] R. Haumann, J. C. Videira, G. J. L. Kaspers, D. G. van Vuurden, and E. Hulleman, "Overview of Current Drug Delivery Methods Across the Blood–Brain Barrier for the Treatment of Primary Brain Tumors," *CNS Drugs*, vol. 34, no. 11, pp. 1121–1131, 2020, doi: 10.1007/s40263-020-00766-w.

[15] Buthayna G Elshaikh, Hiba Omer, MEM Garelnabi, Abdelmoneim Sulieman, Nawader Abdella, Sahar Algadi and Hind Toufig, "Incidence, Diagnosis and Treatment of Brain Tumours," *Article in Journal of Research in Medical and Dental Science*, vol. 9, no. 6, pp. 340–347, 2021, [Online]. Available: https://www.researchgate.net/publication/353757171

[16] MAYO CLINIC, "Glioblastoma." [Online]. Available: https://www.mayoclinic.org/diseases-conditions/glioblastoma/symptoms-causes/syc-20569077

[17] Yasuteru Inoue, Francis Shue, Guojun Bu and Takahisa Kanekiyo, "Pathophysiology and probable etiology of cerebral small vessel disease in vascular dementia and Alzheimer's disease," *Molecular Neurodegeneration*, vol. 18, no. 1, pp. 1–22, 2023, doi: 10.1186/s13024-023-00640-5.

[18] R. Sahathevan, A. Brodtmann, and G. A. Donnan, "Dementia, stroke, and vascular risk factors; a review," *International Journal of Stroke*, vol. 7, no. 1, pp. 61–73, 2012, doi: 10.1111/j.1747-4949.2011.00731.x.

[19] FOUNDATION TO ADVANCE VASCULAR CURES, "Vascular Dementia Image." Accessed: Jul. 01, 2025. [Online]. Available: https://www.vascularcures.org/vascular-dementia

[20] Kateřina Menšíková, Radoslav Matěj, Carlo Colosimo, Raymond Rosales, Lucie Tučková, Jiří Ehrmann, Dominik Hraboš, Kristýna Kolaříková, Radek Vodička, Radek Vrtěl, Martin Procházka, Martin Nevrlý, Michaela Kaiserovál, Sandra Kurčovál, Pavel Otruba and Petr, "Lewy body disease or diseases with Lewy bodies?," *npj Parkinson's Disease*, vol. 8, no. 1, 2022, doi: 10.1038/s41531-021-

00273-9.

[21] Alzheimer's Association, "Dementia with Lewy Bodies." Accessed: Jun. 18, 2025. [Online]. Available: https://www.alz.org/alzheimers-dementia/what-is-dementia/types-of-dementia/dementia-with-lewy-bodies

[22] Indu Bhushan, Manjot Kour, Guneet Kour, Shriya Gupta, Supriya Sharma and Arvind Yadav, "Alzheimer's disease: Causes & treatment – A review," *Annals of Biotechnology*, vol. 1, no. 1, 2018, doi: 10.33582/2637-4927/1002.

[23] M. Safieh, A. D. Korczyn, and D. M. Michaelson, "ApoE4: an emerging therapeutic target for Alzheimer's disease," *BMC Medicine*, vol. 17, no. 1, pp. 1–17, 2019, doi: 10.1186/s12916-019-1299-4.

[24] Prashanthi Vemuri, Timothy G. Lesnick, Scott A. Przybelski, David S. Knopman, Rosebud O. Roberts, Val J. Lowe, Kejal Kantarci, Mathew L. Senjem, Jeffrey L. Gunter, Bradley F. Boeve, Ronald C. Petersen, and Clifford R. Jack Jr, "Effect of lifestyle activities on alzheimer disease biomarkers and cognition," *Annals of Neurology*, vol. 72, no. 5, pp. 730–738, 2012, doi: 10.1002/ana.23665.

[25] U. Jiwtode, S. Chakole, and N. Bhatt, "Alzheimer's Disease: History, Stages, Diagnosis and Its Future," *Journal of Pharmaceutical Research International*, vol. 33, pp. 41–45, 2021, doi: 10.9734/jpri/2021/v33i39a32139.

[26] Clifford Jack Jr, J. Scott Andrews, Thomas G. Beach, Teresa Buracchio, Billy Dunn, Ana Graf, Oskar Hansson, Carole Ho, William Jagust, Eric McDade, Jose Luis Molinuevo, Ozioma C. Okonkwo, Luca Pani, Michael S. Rafii, Philip Scheltens and Eric

Siemers, , "Revised criteria for diagnosis and staging of Alzheimer's disease: Alzheimer's Association Workgroup," *Alzheimer's and Dementia*, vol. 20, no. 8, pp. 5143–5169, 2024, doi: 10.1002/alz.13859.

[27] C. Bruno Dubois, Nicolas Villain, Giovanni B Frisoni, Gil D Rabinovici, Marwan Sabbagh, Stefano Cappa, Alexandre Bejanin, Stéphanie Bombois, Stéphane Epelbaum, Marc Teichmann, Marie-Odile Habert, Agneta Nordberg, Kaj Blennow, Douglas Galasko, Yaakov Stern, "Clinical diagnosis of Alzheimer's disease: recommendations of the International Working Group. Lancet Neurol," *The Lancet Neurology*, 2024.

[28] Alzheimer's Association, "2010 Alzheimer's disease facts and figures.," *Alzheimer's & dementia : the journal of the Alzheimer's Association*, vol. 6, no. 2, pp. 158–194, Mar. 2010, doi: 10.1016/j.jalz.2010.01.009.

[29] H. Alkaales and S. Ibrahim, "Alpha 1- Anti-chymotrypsin and Other Inflammatory Factors As Biomarkers for Alzheimer's Disease," *Egyptian Academic Journal of Biological Sciences. C, Physiology and Molecular Biology*, vol. 15, no. 2, pp. 769–774, 2023, doi: 10.21608/eajbsc.2023.328457.

[30] I. Santana, F. Farinha, S. Freitas, V. Rodrigues, and Á. Carvalho, "Epidemiologia da Demência e da Doença de Alzheimer em Portugal: Estimativas da Prevalência e dos Encargos Financeiros com a Medicação," *Acta Médica Portuguesa*, vol. 28, no. 2, pp. 182–188, 2015, doi: 10.20344/amp.6025.

[31] Luke William Bonham, Desikan Rahul Singh and Yokoyama Jennifer Sue, "The relationship between complement factor C3,

APOE ε4, amyloid and tau in Alzheimer's disease," *Acta neuropathologica communications*, vol. 4, no. 1, p. 65, 2016, doi: 10.1186/s40478-016-0339-y.

[32] M. A. Rahman, M. S. Rahman, M. J. Uddin, A. N. M. Mamum-Or-Rashid, M. G. Pang, and H. Rhim, "Emerging risk of environmental factors: insight mechanisms of Alzheimer's diseases," *Environmental Science and Pollution Research*, vol. 27, no. 36, pp. 44659–44672, 2020, doi: 10.1007/s11356-020-08243-z.

[33] Catherine Reed, Mark Belger, Dell' Agnello Grazia, Wimo Anders, Josep Maria, Argimon, Bruno Giuseppe, Dodel Richard, Josep Maria Haro, Roy W Jones and Vellas Bruno, "Caregiver Burden in Alzheimer's Disease: Differential Associations in Adult-Child and Spousal Caregivers in the GERAS Observational Study," *Dementia and Geriatric Cognitive Disorders Extra*, vol. 4, no. 1, pp. 51–64, 2014, doi: 10.1159/000358234.

[34] Hashima E. Nasreen, Marie Tyrrel, Sofia Vikström, Åsa Craftman, Syarifah Amirah Binti Syed Ahmad, Nora Mat Zin, Karimah Hanim Abd Aziz, Noorlaili Binti Mohd Tohit, Mohd Aznan Md Aris and Zarina Nahar Kabir, "Caregiver burden, mental health, quality of life and self-efficacy of family caregivers of persons with dementia in Malaysia: baseline results of a psychoeducational intervention study," *BMC Geriatrics*, vol. 24, no. 1, pp. 1–12, 2024, doi: 10.1186/s12877-024-05221-9.

[35] Randall James Bateman, Paul Stephen Aisen, Bart De Strooper, Nick Charles Fox, Cynthia Ann Lemere, John Michael Ringman, Stephen Salloway, Reisa Ann Sperling, Manfred Windisch and Chengjie Xiong, "Autosomal-dominant Alzheimer's disease: a review and proposal for the prevention of Alzheimer's disease,"

Alzheimer's Research & Therapy, vol. 3, no. 1, pp. 1–13, 2011, doi: 10.1093/oed/3725923206.

[36] J. Song, J. Zheng, P. Li, X. Lu, G. Zhu, and P. Shen, "An Effective Multimodal Image Fusion Method Using MRI and PET for Alzheimer's Disease Diagnosis," *Frontiers in Digital Health*, vol. 3, pp. 1–12, 2021, doi: 10.3389/fdgth.2021.637386.

[37] Neurotorium, "What factors are at play in Alzheimer's disease." Accessed: May 01, 2025. [Online]. Available: https://neurotorium.org/image/what-factors-are-at-play-in-alzheimers-disease/

[38] He´lène-Marie Lanoisele´, Gae¨ l Nicolas, David Wallon, Anne Rovelet Lecrux, Morgane Lacour, Ste´phane Rousseau, Anne-Claire Richard, Florence Pasquier, Adeline Rollin-Sillaire, Olivier Martinaud, Muriel Quillard-Muraine, Vincent de la Sayette, Claire Bou, "APP, PSEN1, and PSEN2 mutations in early-onset Alzheimer disease: A genetic screening study of familial and sporadic cases," *PLoS Medicine*, vol. 14, no. 3, pp. 1–16, 2017, doi: 10.1371/journal.pmed.1002270.

[39] Felicia Liana Andronie-Cioara, Adriana Ioana Ardelean, Carmen Delia Nistor-Cseppento, Anamaria Jurcau, Maria Carolina Jurcau, Nicoleta Pascalau and Florin Marcu, "Molecular Mechanisms of Neuroinflammation in Aging and Alzheimer's Disease Progression," *International Journal of Molecular Sciences*, vol. 24, no. 3, 2023, doi: 10.3390/ijms24031869.

[40] Michelle M. Mielke, Neelum T. Aggarwal, Clara Vila-Castelar, Puja Agarwal, Eider M.Arenaza-Urquijo, Anna Brugulat-Serrat, Lyndsey E. DuBose, Benjamin Brett, Willem S. Eikelboom, Jason

Flatt, Nancy S. Foldi, Sanne Franzen, Alison J. McManus, Debora Melo va, "Consideration of sex and gender in Alzheimer's disease and related disorders from a global perspective," *Alzheimer's and Dementia*, vol. 18, no. 12, pp. 2707–2724, 2022, doi: 10.1002/alz.12662.

[41] José Alberto Luchsinger and Richard Mayeux, "Dietary factors and Alzheimer's disease José," *Lancet Neurology*, vol. 3, no. 10, pp. 579–587, 2004, doi: 10.1016/S1474-4422(04)00870-2.

[42] N. Korte, R. Nortley, and D. Attwell, "Cerebral blood flow decrease as an early pathological mechanism in Alzheimer's disease," *Acta Neuropathologica*, vol. 140, no. 6, pp. 793–810, 2020, doi: 10.1007/s00401-020-02215-w.

[43] Y. Chen, M. Dang, and Z. Zhang, "Brain mechanisms underlying neuropsychiatric symptoms in Alzheimer's disease: a systematic review of symptom-general and –specific lesion patterns," *Molecular Neurodegeneration*, vol. 16, no. 1, pp. 1–22, 2021, doi: 10.1186/s13024-021-00456-1.

[44] Xiao-Ling Li, Nan Hu, Meng-Shan Tan, Jin-Tai Yu and Lan Tan, "Behavioral and Psychological Symptoms in Alzheimer's Disease," *BioMed Research International*, p. 9, 2014, doi: 10.1155/2014/927804.

[45] A. Starzyk and P. Charzewski, "The Science of Memory Loss : Insights into Alzheimer ' s Disease," *Quality in Sport*, pp. 1–11, 2025.

[46] R. Ben Ammar and Y. Ben Ayed, "Language-related features for early detection of Alzheimer Disease," *Procedia Computer Science*, vol. 176, pp. 763–770, 2020, doi: 10.1016/j.procs.2020.09.071.

[47] H. Chern, G. Caruso, H. Desaire, and R. Jarosova, "Carnosine Mitigates Cognitive Impairment and Dopamine Release in an Okadaic Acid-Induced Zebrafish Model with Alzheimer's Disease-like Symptoms," *ACS Chemical Neuroscience*, 2025.

[48] I. Bazarbekov, A. Razaque, M. Ipalakova, J. Yoo, Z. Assipova, and A. Almisreb, "A review of artificial intelligence methods for Alzheimer's disease diagnosis: Insights from neuroimaging to sensor data analysis," *Biomedical Signal Processing and Control*, vol. 92, p. 106023, 2024.

[49] E. Storey, G. J. Kinsella, and M. J. Slavin, "The neuropsychological diagnosis of Alzheimer's disease," *Journal of Alzheimer's Disease*, vol. 3, no. 3, pp. 261–285, 2001, doi: 10.3233/JAD-2001-3302.

[50] H. M. Zeng, H. B. Han, Q. F. Zhang, and H. Bai, "Application of modern neuroimaging technology in the diagnosis and study of Alzheimer's disease," *Neural Regeneration Research*, vol. 16, no. 1, pp. 73–79, 2021, doi: 10.4103/1673-5374.286957.

[51] W. van der F. and K. B. Stephanie Johanna Beatrix Vos, Ineke Anna van Rossum, Frans Verhey, Dirk Lodewijk Knol, Hilkka Soininen, Lars-Olof Wahlund, Harald Hampel, Magda Tsolaki, Lennart Minthon, Giovanni Battista Frisoni, Lutz Froelich, Flavio Nobili, "Prediction of Alzheimer disease in subjects with amnestic and nonamnestic MCI," *Neurology*, vol. 80, no. 12, pp. 1124–1132, 2013, doi: 10.1212/WNL.0b013e318288690c.

[52] Nadjia Amini, Mounir Ibn Hach, Laurence Lapauw, Jolan Dupont, Laura Vercauteren, Sabine Verschueren, Jos Tournoy and Evelien Gielen, "Meta-analysis on the interrelationship between sarcopenia and mild cognitive impairment, Alzheimer's disease and other

forms of dementia," *Journal of Cachexia, Sarcopenia and Muscle*, vol. 15, no. 4, pp. 1240–1253, 2024, doi: 10.1002/jcsm.13485.

[53] Muazzam Maqsood, Faria Nazir, Umair Khan, Farhan Aadil and Habibullah Jamal, "Transfer Learning Assisted Classification and Detection of Alzheimer's Disease Stages Using 3D MRI Scans," *Sensors*, vol. 19, no. 12, p. 2645, 2019, doi: 10.3390/s19112645.

[54] Farheen Ramzan, Muhammad Usman Ghani Khan, Asim Rehmat, Sajid Iqbal, Tanzila Saba and Amjad Rehman, "A Deep Learning Approach for Automated Diagnosis and Multi-Class Classification of Alzheimer's Disease Stages Using Resting-State fMRI and Residual Neural Networks," *Journal of Medical Systems*, vol. 44, no. 2, pp. 1–16, 2020, doi: 10.1007/s10916-019-1475-2.

[55] Rizwan Khan, Saeed Akbar, Atif Mehmood, Farah Shahid, Khushboo Munir, Naveed Ilyas, Muhammad Asif and Zhonglong Zheng, "A transfer learning approach for multiclass classification of Alzheimer's disease using MRI images," *Frontiers in Neuroscience*, vol. 16, p. 1050777, 2023, doi: 10.3389/fnins.2022.1050777.

[56] Martin Dyrba, Michael Ewers, Martin Wegrzyn, Ingo Kilimann, Claudia Plant, Annahita Oswald, Thomas Meindl, Michela Pievani, Arun Lakshmanan, Andreas Fellgiebel, Massimo Filippi, Harald Hampel, Stefan Klöppel, Karlheinz Hauenstein, Thomas Kirste and Stefan, "Robust Automated Detection of Microstructural White Matter Degeneration in Alzheimer's Disease Using Machine Learning Classification of Multicenter DTI Data," *PLoS ONE*, vol. 8, no. 5, p. e64925, 2013, doi: 10.1371/journal.pone.0064925.

[57] Stephanie Johanna Beatrix Vos, Ineke Anna van Rossum, Frans Verhey, Dirk Lodewijk Knol, Hilkka Soininen, Lars-Olof Wahlund,

Harald Hampel, Magda Tsolaki, Lennart Minthon, Giovanni Battista Frisoni, Lutz Froelich, Flavio Nobili, Wiesje van der Flier and Ka, "Prediction of Alzheimer disease in subjects with amnestic and nonamnestic MCI," *Neurology*, vol. 80, no. 12, pp. 1124–1132, 2013, doi: 10.1212/WNL.0b013e318288690c.

[58] Rajaram Sampath and Arunachalam Saradha, "Classification of Alzheimer's Disease Stages Exploiting an ANFIS Classifier," *International Journal of Applied Engineering Research*, vol. 9, no. 22, pp. 16979–16990, 2014, [Online]. Available: http://www.ripublication.com

[59] Giulia Fiscon, Emanuel Weitschek, Giovanni Felici, Paola Bertolazzi, Simona De Salvo, Placido Bramanti and Maria Cristina De Cola, "Alzheimer's disease patients classification through EEG signals processing," *Symposium on Computational Intelligence and Data Mining, Proceeding*, pp. 105–112, 2014, doi: 10.1109/CIDM.2014.7008655.

[60] Ayşe Demirhan, Talia Miriam Nir, Artemis Zavaliangos Petropulu, Clifford Ray Jack Jr, Michael William Weiner, Matt Alan Bernstein, Paul Michael Thompson, Neda Jahanshad, and the Alzheimer's Disease Neuroimaging Initiative (ADNI), "FEATURE SELECTION IMPROVES THE ACCURACY OF CLASSIFYING ALZHEIMER DISEASE USING DIFFUSION TENSOR IMAGES," *Proceedings of the IEEE International Symposium on Biomedical Imaging*, vol. 176, no. 1, pp. 126–130, 2015, doi: 10.1109/ISBI.2015.7163832.FEATURE.

[61] Hamed Taheri Gorji and Javad Haddadnia, "A novel method for early diagnosis of Alzheimer's disease based on pseudo Zernike moment from structural MRI," *Neuroscience*, vol. 305, pp. 361–

371, 2015, doi: 10.1016/j.neuroscience.2015.08.013.

[62] Saman Sarraf and Ghassem Tofighi, "Classification of Alzheimer's Disease Structural MRI Data by Deep Learning Convolutional Neural Networks," *arXiv preprint*, 2016, doi: arXiv:1603.08631v1.

[63] Debesh Jha and Goo-Rak Kwon, "Alzheimer Disease Detection in MRI Using Curvelet Transform with K-NN," *Journal of Korean Institute of Information Technology*, vol. 14, no. 8, pp. 121–129, 2016, doi: 10.14801/jkiit.2016.14.8.121.

[64] Marcia Hon and Naimul Mefraz Khan, "Towards Alzheimer's Disease Classification through Transfer Learning," *arXiv:1711.11117v1*, 2017.

[65] Kajal Kiran Gulhare, Soorya Prakash Shukla and Lokesh Kumar Sharma, "Deep Neural Network Classification method to Alzheimer's Disease Detection," *International Journals of Advanced Research in Computer Science and Software Engineering*, vol. 7, no. 6, 2017, doi: 10.13140/RG.2.2.31207.75687.

[66] Jyoti Islam and Yanqing Zhang, "Brain MRI analysis for Alzheimer's disease diagnosis using an ensemble system of deep convolutional neural networks," *Brain Informatics*, vol. 5, no. 2, 2018, doi: 10.1186/s40708-018-0080-3.

[67] Alexander Khvostikov, Karim Aderghal, Jenny Benois-Pineau, Andrey Krylov and Gwenaelle Catheline, "3D CNN-based classification using sMRI and MD-DTI images for Alzheimer disease studies," *arXiv:1801.05968v1*, vol. 1, 2018, doi: 10.1016/j.neuroimage.2018.01.070.

[68] Firouzeh Razavi, Mohammad Jafar Tarokh and Mahmood Alborzi,

"An intelligent Alzheimer's disease diagnosis method using unsupervised feature learning," *Journal of Big Data*, vol. 6, no. 32, 2019, doi: 10.1186/s40537-019-0190-7.

[69] Sameer Mahajan, Gayatri Bangar and Nahush Kulkarni, "Machine Learning Algorithms for Classification of Various Stages of Alzheimer's Disease," *International Research Journal of Engineering and Technology (IRJET)*, vol. 07, no. 08, pp. 817–822, 2020.

[70] Akash Rajak and Ajay Kumar Shrivastava, "Diagnosis of Alzheimer Disease using Machine Learning Approaches," *International Journal of Advanced Science and Technology*, vol. 29, no. 04, pp. 7062–7073, 2020, [Online]. Available: https://www.researchgate.net/publication/342847161

[71] Nivedhitha Mahendran, Ponnusamy Mahendran Durai Raj Vincent, Kathiravan Srinivasan and Chuan-Yu Chang, "Improving the Classification of Alzheimer's Disease Using Hybrid Gene Selection Pipeline and Deep Learning," *Frontiers in Genetics*, vol. 12, p. 784814, 2021, doi: 10.3389/fgene.2021.784814.

[72] Xiaowang Bi, Wei Liu, Huaiqin Liu and Qun Shang, "Artificial Intelligence-basedMRIImages for Brain in Prediction of Alzheimer's Disease," *Journal of Healthcare Engineering*, 2021, doi: 10.1155/2021/8198552.

[73] Monika Sethi, Shalli Rani, Aman Singh and Juan Luis Vidal Mazón, "A CAD System for Alzheimer's Disease Classification Using Neuroimaging MRI 2D Slices," *Computational and Mathematical Methods in Medicine*, vol. 2022, 2022, doi: 10.1155/2022/8680737.

[74] Jinhua Sheng, Yu Xin, Qiao Zhang, Luyun Wang, Ze Yang and Jie

Yin, “Predictive classification of Alzheimer’s disease using brain imaging and genetic data,” *Scientific Reports*, vol. 12, no. 1, p. 2405, 2022, doi: 10.1038/s41598-022-06444-9.

[75] Mohi ud din dar,Avinash Bhagat, Syed Immamul Ansarullah, Mohamed Tahar Ben Othman, Yasir Hamid, Hend Khalid Alkahtani, Inam Ullah and Habib Hamam, “A Novel Framework for Classification of Different Alzheimer’s Disease Stages Using CNN Model,” *Electronics*, vol. 12, no. 2, p. 469, 2023, doi: 10.3390/electronics12020469.

[76] Sheng Chen, Chutian Zhang, Hongjun Yang, Liang Peng, Haiqun Xie, Zeping Lv and Zeng-Guang Hou, “A Multi-Modal Classification Method for Early Diagnosis of Mild Cognitive Impairment and Alzheimer’s Disease Using Three Paradigms With Various Task Difficulties,” *IEEE Transactions on Neural Systems and Rehabilitation Engineering*, vol. 32, pp. 1477–1483, 2024, doi: 10.1109/TNSRE.2024.3379891.

[77] Subhrangshu Bit, Pritam Dey, Arnab Maji and Tapan K Khan, “MRI-based mild cognitive impairment and Alzheimer’s disease classification using an algorithm of combination of variational autoencoder and other machine learning classifier,” *Journal of Alzheimer’s Disease Reports*, vol. 8, pp. 1093–1111, 2024, doi: 10.1177/25424823241290694.

[78] Laith Alzubaidi, Juan Santamaría, Mohammed Ali Fadhel, Muthana Al-Amidie, Laith Farhan, Jinglan Zhang, Amjad J. Humaidi, Ayad Al-Dujaili, Ye Duan and Omran Al-Shamma, “Review of deep learning: concepts, CNN architectures, challenges, applications, future directions,” *Journal of Big Data*, vol. 8, no. 1, 2021, doi: 10.1186/s40537-021-00444-8.

[79] Yann LeCun, Yoshua Bengio and Geoffrey Hinton, "Deep learning," *Nature*, vol. 29, no. 7553, pp. 1–73, 2016, [Online]. Available: http://deeplearning.net/

[80] Mohamed Ahzam Amanullah, Riyaz Ahamed Ariyaluran Habeeb, Fariza Hanum Nasaruddin, Abdullah Gani, Ejaz Ahmed, Abdul Salam Mohamed Nainar, Nazihah Md Akim, Muhammad Imran, "Deep learning and big data technologies for IoT security," *Computer Communications*, vol. 151, pp. 495–517, 2020, doi: 10.1016/j.comcom.2020.01.016.

[81] Amey Thakur and Archit Konde, "Fundamentals of Neural Networks," *International Journal for Research in Applied Science and Engineering Technology*, vol. 9, no. VIII, pp. 407–426, 2021, doi: 10.22214/ijraset.2021.37362.

[82] M. Rawat, H. Khandelwal, and N. Gupta, "The Essentials of Neural Networks and their Applications," *Neural networks*, vol. 7, pp. 371–375, 2020.

[83] Raid Rafi Omar Al-Nima, "Signal Processing and Machine Learning Techniques for Human Verification Based on Finger Textures," PhD thesis, School of Engineering, Newcastle University, UK, 2017.

[84] Marwa MohamedSheet AL-Hatab, Raid Rafi Omar Al-Nima, Ilaria Marcantoni, Camillo Porcaro, Laura Burattini, "Classifying Various Brain Activities by Exploiting Deep Learning Techniques and Genetic Algorithm Fusion Method," *TEST Engineering & Management*, vol. 83, no. 3035, pp. 3035–3052, 2020.

[85] M. M. AL-Hatab, R. R. O. Al-Nima, I. Marcantoni, C. Porcaro, and L. Burattini, "Comparison Study Between Three Axis Views of

Vision, Motor and Pre-Frontal Brain Activities," *Journal of Critical Reviews*, vol. 7, no. 5, pp. 2598–2607, 2020.

[86] S. Mascarenhas and M. Agarwal, "A comparison between VGG16, VGG19 and ResNet50 architecture frameworks for Image Classification," in *2021 International conference on disruptive technologies for multi-disciplinary research and applications (CENTCON)*, IEEE, 2021, pp. 96–99.

[87] G. Shamsipour, S. Fekri-Ershad, M. Sharifi, and A. Alaei, "Improve the efficiency of handcrafted features in image retrieval by adding selected feature generating layers of deep convolutional neural networks," *Signal, image and video processing*, vol. 18, no. 3, pp. 2607–2620, 2024.

[88] Muhathir, M. F. Dwi Ryandra, R. B. Y. Syah, N. Khairina, and R. Muliono, "Convolutional neural network (CNN) of Resnet-50 with Inceptionv3 architecture in classification on X-ray image," in *Computer Science On-line Conference*, Springer, 2023, pp. 208–221.

[89] A. A. Almisreb, N. Jamil, and N. M. Din, "Utilizing AlexNet Deep Transfer Learning for Ear Recognition," *Proceedings - 2018 4th International Conference on Information Retrieval and Knowledge Management: Diving into Data Sciences, CAMP 2018*, no. April 2019, pp. 8–12, 2018, doi: 10.1109/INFRKM.2018.8464769.

[90] S. A. Albelwi, "Deep Architecture based on DenseNet-121 Model for Weather Image Recognition," *International Journal of Advanced Computer Science and Applications*, vol. 13, no. 10, pp. 559–565, 2022, doi: 10.14569/IJACSA.2022.0131065.

[91] Bijen Khagi, Chung Ghiu Lee and Goo-Rak Kwon, "Alzheimer's

disease classification from brain MRI based on transfer learning from CNN," in *biomedical engineering international conference (BMEiCON)*, IEEE, 2018.

[92] Mahbub Hussain, Jordan James Bird and Faria Diego Ricardo, "A Study on CNN Transfer Learning for Image Classificatio," *Advances in Intelligent Systems and Computing*, vol. 840, pp. 191–202, 2019, doi: 10.1007/978-3-319-97982-3_16.

[93] Alex Krizhevsky, Ilya Sutskever and Geoffrey E. Hinton, "Neural Networks," *Encyclopedia of Movement Disorders, Three-Volume Set*, vol. 3, pp. V2-257-V2-259, 2010, doi: 10.1016/B978-0-12-374105-9.00493-7.

[94] T. A. M. and S. A.-Z. Saad Albawi, "Understanding of a Convolutional Neural Network," *Proceedings of 2017 International Conference on Engineering and Technology*, pp. 1–6, 2017, doi: 10.1109/ICEngTechnol.2017.8308186.

[95] Ahmed Saeed Ibrahim Al-Obaidi, Raid Rafi Omar Al-Nima, and Tingting Han, "Interpreting Arabic Sign Alphabet by Using the Deep Learning," *AIP Conference Proceedings*, vol. 2862, no. 1, pp. 29–30, 2023, doi: 10.1063/5.0171428.

[96] Ihsan Evanjelist Firdausi, Irfan Mukhlash, Atyah Dhamar Syah Gama and Naufal Hidayat, "Sentiment analysis of customer response of telecommunication operator in Twitter using DCNN-SVM Algorithm," *Journal of Physics: Conference Series*, vol. 1490, no. 1, 2020, doi: 10.1088/1742-6596/1490/1/012071.

[97] Ernest Jeczmionek and Piotr Adam Kowalski, "Flattening layer pruning in convolutional neural networks," *Symmetry*, vol. 13, no. 7, pp. 1–13, 2021, doi: 10.3390/sym13071147.

[98] N. Ketkar and J. Moolayil, *Deep Learning with Python*. 2021. doi: 10.1007/978-1-4842-5364-9.

[99] Wu Hao and Zhao Jinsong, "Deep convolutional neural network model based chemical process fault diagnosis," *Computers & chemical engineering*, vol. 115, pp. 185–197, 2018.

[100] Sagheer Abbas , Yousef Alhwaiti , Areej Fatima , Muhammad A. Khan , Muhammad Adnan Khan, Taher M. Ghazal, Asma Kanwal, Munir Ahmad and Nouh Sabri Elmitwally, "Convolutional neural network based intelligent handwritten document recognition," *Computers, Materials and Continua*, vol. 70, no. 3, pp. 4563–4581, 2022, doi: 10.32604/cmc.2022.021102.

[101] Weiyang Liu, Yandong Wen, Zhiding Yu and Meng Yang, "Large-Margin Softmax Loss for Convolutional Neural Networks," *Proceedings of the 33rd International Conference on Machine Learning*, vol. 48, 2016, [Online]. Available: http://arxiv.org/abs/1612.02295

[102] M. Wojciuk, Z. Swiderska-Chadaj, K. Siwek, and A. Gertych, "Improving classification accuracy of fine-tuned CNN models: Impact of hyperparameter optimization," *Heliyon*, vol. 10, no. 5, p. e26586, 2024, doi: 10.1016/j.heliyon.2024.e26586.

[103] M. Dhouibi, A. K. Ben Salem, and S. Ben Saoud, "Optimization of CNN model for image classification," *3rd IEEE International Conference on Design and Test of Integrated Micro and Nano-Systems, DTS 2021*, no. June, 2021, doi: 10.1109/DTS52014.2021.9497988.

[104] H. Lee and J. Song, "Introduction to convolutional neural network using Keras; An understanding from a statistician,"

Communications for Statistical Applications and Methods, vol. 26, no. 6, pp. 591–610, 2019, doi: 10.29220/CSAM.2019.26.6.591.

[105] Mohamad Zaim Awang Pon And Krishna Prakash, "Hyperparameter Tuning of Deep learning Models in Keras," *Sparklinglight Transactions on Artificial Intelligence and Quantum Computing*, vol. 1, no. 1, pp. 23–35, 2021.

[106] S. Roy, R. Mehera, R. K. Pal, and S. K. Bandyopadhyay, "Hyperparameter optimization for deep neural network models: a comprehensive study on methods and techniques," *Innovations in Systems and Software Engineering*, pp. 1–12, 2023.

[107] Samuel Gbashi, Obafemi Olatunji, Paul Adedeji and Nkosinathi Madushele, "Hyperparameter Optimization on CNN Using Hyperband for fault identification in wind turbine high-speed shaft gearbox bearing," *International Conference on Electrical, Computer and Energy Technologies, ICECET 2023*, no. November, pp. 1–7, 2023, doi: 10.1109/ICECET58911.2023.10389387.

[108] Rana Ahmed Lateef and Ayad Razi Abbas, "Tuning the Hyperparameters of the 1D CNN Model to Improve the Performance of Human Activity Recognition," *Engineering and Technology Journal*, vol. 40, no. 4, pp. 547–554, 2022, doi: 10.30684/etj.v40i4.2054.

[109] T. Akiba, S. Sano, T. Yanase, T. Ohta, and M. Koyama, "Optuna: A Next-generation Hyperparameter Optimization Framework," *Proceedings of the ACM SIGKDD International Conference on Knowledge Discovery and Data Mining*, pp. 2623–2631, 2019, doi: 10.1145/3292500.3330701.

[110] Yoshihiko Ozaki, Yuki Tanigaki, Shuhei Watanabe, Masahiro

Nomura and Masaki Onishi, "Multiobjective Tree-Structured Parzen Estimator," *Journal of Artificial Intelligence Research*, vol. 73, pp. 1209–1250, 2022, doi: 10.1613/jair.1.13188.

[111] Jenia Afrin Jeba, "Case study of Hyperparameter optimization framework Optuna on a Multi-column Convolutional Neural Network," University of Saskatchewan, Saskatoon, Canada, 2021.

[112] Ming-Jr Tsai, Chung-Hui Lin, Jung-Pin Lai and Ping-Feng Pai, "Using Deep Learning, Optuna, and Digital Images to Identify Necrotizing Fasciitis," *Electronics (Switzerland)*, vol. 13, no. 22, pp. 1–16, 2024, doi: 10.3390/electronics13224421.

[113] RP5, "Rasbperry Pi 5," Accessed May , 2025. [Online]. Available: https://www.raspberrypi.com/products/raspberry-pi-5/

[114] Marina Sokolova, Nathalie Japkowicz and Stan Szpakowicz, "Beyond Accuracy, F-score and ROC: a Family of Discriminant Measures for Performance Evaluation," *Advances in Artificial Intelligence*, pp. 1015–1021, 2006.

[115] S. S. Martínez, D. M. Gila, A. Beyaz, J. G. Ortega, and J. G. García, "A computer vision approach based on endocarp features for the identification of olive cultivars," *Computers and electronics in agriculture*, vol. 154, pp. 341–346, 2018.

[116] N. Saqib, K. F. Haque, V. P. Yanambaka, and A. Abdelgawad, "Convolutional-Neural-Network-Based Handwritten Character Recognition: An Approach with Massive Multisource Data," *Algorithms*, vol. 15, no. 4, 2022, doi: 10.3390/a15040129.

[117] Hossam Magdy Balaha, Hesham Arafat Ali, Mohamed Saraya and Mahmoud Badawy, "A new Arabic handwritten character recognition deep learning system (AHCR-DLS)," *Neural*

Computing and Applications, vol. 33, no. 11, pp. 6325–6367, 2021, doi: 10.1007/s00521-020-05397-2.

[118] D. Sanyal, N. Bosch, and L. Paquette, "Feature Selection Metrics: Similarities, Differences, and Characteristics of the Selected Models," *Proceedings of the 13th International Conference on Educational Data Mining, EDM 2020*, pp. 212–223, 2020.

[119] Kaggle, "Best Alzheimer MRI dataset," *Accessed September 30, 2024*, 2024, [Online]. Available: https://www.kaggle.com/datasets/lukechugh/best-alzheimer-mri-dataset-99-accuracy

[120] Kaggle, "Alzheimers disease MRI images," *Accessed September 30, 2024*, 2024, [Online]. Available: https://www.kaggle.com/datasets/shariaarfin/alzheimers-disease-mri-images?resource=download

Kolibri 360 GmbH
Ferdinand-Jühlke-Straße 7
99095 Erfurt, Deutschland
produktsicherheit@kolibri360.de

Druck:
CPI Druckdienstleistungen GmbH
im Auftrag der
Kolibri 360 GmbH
Ferdinand-Jühlke-Str. 7
99095 Erfurt